A Practical Guide to Leading Green Schools

This practical guide for educational leaders explores how you can transform your school or district into a vibrant center of learning and socio-ecological responsibility with only three manageable actions: taking students outside, bringing nature inside, and cultivating a mindset of awareness, responsibility, and empathy. This book is rich in practical, attainable approaches and stories of real actions taken by leaders, teachers, parents, and community partners to design, lead, and manage a vibrant, flourishing, sustainable learning community. Authors Uline and Kensler take you on an inspirational journey through nine key leadership strategies for you to begin or expand your work towards whole school sustainability.

Cynthia L. Uline is former Director of the National Center for the 21st Century Schoolhouse and Professor Emeritus of Educational Leadership at San Diego State University, USA.

Lisa A. W. Kensler is the Emily R. and Gerald S. Leischuck Endowed Professor of Educational Leadership at Auburn University, USA.

Other Eye On Education Books Available from Routledge
(www.routledge.com/eyeoneducation)

A Practical Guide to Leading Green Schools

Partnering with Nature to Create Vibrant, Flourishing, Sustainable Schools

Cynthia L. Uline and
Lisa A. W. Kensler

Routledge
Taylor & Francis Group

NEW YORK AND LONDON

First published 2021
by Routledge
52 Vanderbilt Avenue, New York, NY 10017

and by Routledge
2 Park Square, Milton Park, Abingdon, Oxon, OX14 4RN

Routledge is an imprint of the Taylor & Francis Group, an informa business

© 2021 Taylor & Francis

Library of Congress Cataloging-in-Publication Data
Names: Uline, Cynthia L., author. | Kensler, Lisa A. W., author.
Title: A practical guide to leading green schools : partnering with nature to create vibrant, flourishing, sustainable schools / Cynthia L. Uline, Lisa A.W. Kensler.
Identifiers: LCCN 2020054092 (print) | LCCN 2020054093 (ebook) | ISBN 9780367419400 (hardback) | ISBN 9780367422639 (paperback) | ISBN 9780367823016 (ebook)
Subjects: LCSH: Environmental education. | Schools—Environmental aspects. | School improvement programs. | School management and organization.
Classification: LCC GE70 .U45 2021 (print) | LCC GE70 (ebook) | DDC 333.7071—dc23
LC record available at https://lccn.loc.gov/2020054092
LC ebook record available at https://lccn.loc.gov/2020054093

ISBN: 978-0-367-41940-0 (hbk)
ISBN: 978-0-367-42263-9 (pbk)
ISBN: 978-0-367-82301-6 (ebk)

Typeset in Optima
by Apex CoVantage, LLC

For our precious grandchildren, Finn, Alice, Emily, and Alexis, along with all their peers worldwide, who deserve vibrant, flourishing, sustainable schools today and a healthy planet for all their tomorrows.

Contents

Preface

This book serves as a practice-oriented companion to our recent book, *Leadership for Green Schools: Sustainability for Our Children, Our Communities, and Our Planet*, published by Routledge in 2017. This companion book focuses on three straightforward, manageable actions living systems-minded school leaders can implement to move their schools in the direction of whole school sustainability, including:

- **Taking students outside**,
- **Bringing nature inside**, and
- **CARE-ing**, by **C**ultivating and modeling **A**wareness, **R**esponsibility, and **E**mpathy.

Through deep investigation of the possibilities contained within these three actions, as revealed through one particular school district's story, the book demonstrates how pursuit of these actions can transform a school/school district into a vibrant center of learning and socio-ecological responsibility. The featured school district, Encinitas Union School District (EUSD), located in Encinitas, California, was winner of the Green Ribbon School Award from the U.S Department of Education in 2014. The book describes nine specific strategies undertaken by EUSD leaders, teachers, parents, and community partners to reduce their environmental impact and costs, while improving the health and wellness of schools, students, and staff and providing effective environmental and sustainability-focused education. Education for sustainability requires students to stretch for higher, more complex levels of subject matter thinking as they explore and take action on interdependent issues across ecology, society, economy, and well-being domains. Education for sustainability requires teachers accomplish heightened levels of integration across content area and grade levels. Education for sustainability addresses the knowledge and skills necessary for college and careers, from the primary grades on. In this book, we demonstrate how teachers in EUSD, and in other green schools we have studied, engage in close examination of their practice in order to discern if they are challenging their students to high levels of engagement with rigorous, integrated, and context-embedded curricula.

Following an introductory chapter, the remaining nine chapters are divided into three sections, corresponding to the actions presented above. Each section contains three associated strategies, described in detail and illustrated through specific practice-based examples, from Encinitas School District and other Living Systems-Minded Trailblazers throughout the United States. Every chapter is filled with sustainability-related strategies leaders could implement immediately or with little preparation. By spotlighting one school district's story, a journey from SCRAP (a Separate, Compost, Reduce, And Protect cart introduced to students during lunchtime) to whole district sustainability, we provide a reference point, walking the reader through a process of discovering what such change might look like for their school or district.

Special book features include stories of actions taken by *Living Systems-Minded Trailblazers* from across the United States who, like the leaders, teachers, students, and community members in the Encinitas School District, are partnering with nature in numerous ways to design, lead, and manage vibrant, flourishing, sustainable learning communities. Each chapter also closes with a *Leadership Design Challenge*(s). These practical and attainable actions provide school leaders opportunities to, in the words of Encinitas Union School District Superintendent Dr. Timothy Baird, "[not] overthink it, just start it."

The target audience for the book includes practicing and emerging educational leaders, school district administrators, teachers, staff members, students, parents, and community members, all viewed as educational leaders who come together to determine where the entry points for their transformation reside, based on their own school communities' unique needs and circumstances. The book addresses critical social, ecological, and educational challenges of our time. Examples of effective practices, introduced within the context of one school district's lived experience, provide readers guidance for getting started and leading transformative change. The book is written specifically for practicing school leaders in a highly accessible format. Together, our two books demonstrate that leadership for green, sustainability-focused schools is not just an add-on practice, but, rather, the vehicle for 21st-century best practice.

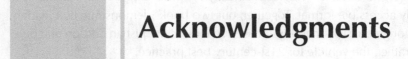

Acknowledgments

Books develop from initial concept to completion with support from so many; many more than we will be able to directly name here. We are honored to have witnessed continued progress and expansion of the green schools movement over the past decade. So many of you in this movement, to bring sustainability related practices to PK–12 schools, have engaged us in conversation, inspired us to act, and supported our work. We acknowledge and appreciate each and every one of you. We hope this book offers you yet another tool for bringing more people, schools, and communities into the work of cultivating vibrant, flourishing, sustainable schools.

We wish to acknowledge our editor, Heather Jarrow, for her resolute belief in this work and her steady hand in shepherding its realization. Anisa Hemming, Director of the Center for Green Schools at the U.S. Green Building Council, sowed the seeds for this book years ago when she challenged us to inspire change by sharing a story that illustrates possibilities turned realities. We acknowledge Anisa's inspiration for this book and her, and her team's, practical support of our research over the past decade. As we imagined early versions of this book, we looked to our friend Jenny Seydel, Executive Director of the Green Schools National Network, for insight into which district leaders were truly leading innovation for sustainability in their school districts. We acknowledge Jenny's important role in introducing us to Dr. Timothy Baird, then Superintendent of Encinitas Union School District (EUSD).

With selection of an exemplary district, we then asked for a great deal of time from EUSD school and district leaders. Kristine Beverly, thank you for aiding our communication and scheduling efforts! We wish to acknowledge everyone's time, passion, and commitment to cultivating vibrant, flourishing, sustainable spaces within which students' love of learning thrives. It was a joy and inspiration to spend so much time in their schools, listening to their stories, and observing their students in action. In particular, Timothy Baird, Andrée Grey, Amy Illingsworth, and Julie Burton, we so appreciate your energy, enthusiasm, and generosity; we could not have written this book without your full engagement and assistance.

We acknowledge and appreciate the support we receive from our friends and colleagues in our home institutions, San Diego State University and Auburn University. We continue to be inspired by individual and collective efforts towards developing future educators and educational leaders for a promising future. We also acknowledge our many friends and colleagues from the University Council for Educational Administration and the American Educational Research Association, who we connect with over conference presentations and conversations. Your scholarship stretches our thinking and ultimately makes our work better.

There are not enough words of gratitude to fully acknowledge our partners, Joe and Mike. They have contributed to this project in so very many tangible and intangible ways, from completing basic chores, to providing inspired moral support, to serving as conceptual sounding boards. Our hearts overflow with love and appreciation for these two.

We join in expressing our deep gratitude for everyone working on behalf of improving schools for children, communities, and planet Earth. In this book, we have highlighted just a few trailblazers beyond EUSD and acknowledge the many more who are leading the way towards healthy ecosystems, just societies, equitable economies, and individual well-being. May we all benefit from getting outside, bringing nature inside, and caring, cultivating awareness, responsibility, and empathy.

Meet the Authors

Lisa A. W. Kensler is the Emily R. and Gerald S. Leischuck Endowed Professor of Educational Leadership in the College of Education at Auburn University. Her original training in ecology continues to fuel her love of applying systems thinking to the challenges located at the intersection of human and nature's systems, particularly as they appear in PK–12 schooling. She has engaged in learning and teaching about systems thinking and sustainability for more than two decades. Lisa's research over the past decade has focused on green schools and the leadership and learning required for transforming schools into more socially just, ecologically healthy, and economically viable communities that engage intentionally with the global sustainability movement. She has published peer-reviewed articles and book chapters related to democratic community, systems thinking, trust, teacher leadership, and whole school sustainability. In 2017, she and Cynthia L. Uline co-authored *Leadership for Green Schools: Sustainability for Our Children, Our Communities, and Our Planet*. In 2018, the University Council for Educational Administration (UCEA) recognized Lisa as one of its Hidden Figures—"behind the scenes giants in the field whose work cannot be ignored."

Cynthia L. Uline is Professor Emeritus of Educational Leadership at San Diego State University and former Director of SDSU's *National Center for the 21st Century Schoolhouse*. Cynthia has also served as a classroom teacher, teacher leader, state education agency administrator, and educational consultant working with school districts, community groups, city governments, state agencies, and governors' offices, always seeking to facilitate meaningful partnerships on behalf of students and their families. For the past 25 years, Cynthia has studied the ways built learning environments support students' learning, as well as the roles leaders, teachers, and community members play in creating learner-centered school facilities. Over the past decade, her research has explored green schools as healthy, vibrant, equitable, and environmentally responsible places for learning. She has published peer-reviewed journal articles and book chapters related to leadership for learning, leadership preparation, whole school sustainability, and the improvement of social and physical learning environments. In 2017, she and Lisa A. W. Kensler co-authored *Leadership for Green Schools: Sustainability for Our Children, Our Communities, and Our Planet* was published by Routledge/Taylor and Francis Group in 2017. She is also a co-author of *Leadership in America's Best Urban Schools* and *Teaching Practices from America's Best Urban Schools, 1st and 2nd Editions.*

Introduction

This is a book about transforming schools, as we know them. Across these pages, we present a leadership design challenge, revealed through the story of a public school district where leaders, teachers, students, parents, and community members partnered with nature in numerous ways to design, lead, and manage a vibrant, flourishing, sustainable learning community. Together, the members of this school community stepped beyond more traditional models of schooling to embrace an alternative aligned with, and reflective of, living systems. When we take time to observe and reflect, we can see how schools already exist as living systems. Each of us is a living system, and we all depend upon the natural systems in which we live. By extension, we are wise to design, manage, and lead our schools with this understanding.

This book has its roots in our previous book, *Leadership for Green Schools: Sustainability for Our Children, Our Communities, and Our Planet*, published by Routledge/Taylor and Francis Group in 2017. Our earlier book was grounded in our personal research into green schools across the United States, as well as a few beyond our borders. The book presented a research-based argument for green schools and for whole school sustainability (WSS), practiced in green schools across the world. We reviewed research across the disciplines of education, psychology, neuroscience, organizational studies, building sciences, ecology, and more. We also described the theoretical principles underlying whole school sustainability as a comprehensive strategy for school improvement, addressing every aspect of education from school culture and climate to curriculum and facilities. As a hands-on companion to *Leadership for Green Schools*, this book, entitled *A Practical Guide to Leading Green Schools: Partnering With Nature to Create Vibrant, Flourishing, Sustainable Schools*, presents a set of key leadership strategies for getting started with the work of whole school sustainability. The target audience for the book includes practicing and emerging educational leaders, school district administrators, teachers, staff members, students, parents, and community members, all viewed as educational leaders who come together to determine where the entry points for their transformation reside, based on their own school communities' unique needs and circumstances.

Introducing Encinitas Union School District

Encinitas Union School District's story, a journey from SCRAP to whole district sustainability, serves as a reference point, walking the reader through a process of discovering what such a transformation might look like for their school and/or school district. This model of schooling, often called whole school or whole district sustainability, integrates sustainability into all aspects of a school organization (Barr, Cross, & Dunbar, 2014), presenting many opportunities for dramatically changing the way schools live, both within their walls and campuses *and* within the larger community and the wider world. From building maintenance to curriculum and instruction, whole school sustainability (WSS) applies living systems understandings to every aspect of school life.

Founded in 1883, the Encinitas Union School District (EUSD), located in north coastal San Diego County, enrolls approximately 5,400 students, housed in nine kindergarten-through-sixth-grade schools and one special education pre-school program. The District serves a diverse and varied community, with a student population that is approximately 68% White, 22% Hispanic, 4% Asian, and 6% other minorities.

All nine schools have been recognized as California Distinguished Schools by the California Department of Education and four have been named National Blue Ribbon Schools by the U.S. Department of Education.

In addition, EUSD was one of nine school districts in the country to receive the Green Ribbon School Award from the U.S Department of Education in April, 2014 for reducing environmental impact and costs, while improving the health and wellness of schools, students, and staff and providing effective environmental and sustainability education. Over the past six years, 64 additional districts have earned this distinction, for a total of 73 out of 13,584 school districts, or one half of 1% of school districts nationwide (https://nces.ed.gov/programs/digest/d17/tables/dt17_214.10. asp). EUSD began their transformation with a simple recycling cart, introduced to students during lunchtime. With this cart, the leaders and teachers across this district sought to make a small difference in their community, reducing landfill waste, facilitating composting, and encouraging lunchtime recycling, at the same time teaching their students about their roles as stewards of the planet. According to one principal,

> Anyone could do a SCRAP cart (Separate, Compost, Reduce, And Protect) and that's where we started. We developed the SCRAP Cart to teach students how to properly sort their lunchtime waste for composting, recycling, and landfill. We began with a pilot, and determined how much trash was being picked up on average. From this baseline, we were able to collect and report data showing the resulting waste reduction. People were floored! We had realized over an eighty percent reduction in waste.

Close consideration of the actions taken by school leaders within the Encinitas Union School District (EUSD) provides opportunities to learn how whole school sustainability can better serve children's well-being and learning, as well as local and global environmental, social, and economic needs in the 21st century. Over the past 11 years, under the leadership of Superintendents Dr. Timothy Baird and Dr. Andrée Grey (who became superintendent following Dr. Baird's retirement in Fall 2019 after serving as Assistant Superintendent of Educational Services for EUSD since 2016), EUSD educators have learned to understand their school/school district as a living system, expanding their view to include the ecological systems upon which their school community depends for clean air, water, food, etc.

Within this living systems context, students thrive. In 2019, overall student performance within the Encinitas School District exceeded state averages by significant margins in both language arts and mathematics. Performance in language arts on the California Assessment of Student Performance and Progress (CAASPP), for students in grades 3–6, stood at 74.79% proficient, exceeding the statewide average of 51.10%. In mathematics, students in grades 3–6 scored 70.67% proficient, as compared with 39.73% statewide. School and district leaders celebrate these successes; at the same time, they are also quick to acknowledge gaps in achievement for three specific subgroups of students, including English learners, low-income students, and students with disabilities. Although the academic performance of students in these subgroups exceeded the state-wide averages, school and district leaders took explicit steps to address these learning challenges, including the creation of English Language Development (ELD) Task Force, the addition of an ELD teacher on Special Assignment, increased teacher professional development opportunities targeting the needs of these student groups, and additional structures and processes to increase cross collaboration between general education and special education. Leaders across the district demonstrate their deep commitment to improving the learning and life outcomes of all their students. This commitment drives them to embrace an expansive, whole systems approach to their work. For example, some schools in the district experience persistent levels of transience and absenteeism among their newcomer immigrant students. A principal in the district described the situation and the school's response.

> This year we've had a really large influx of students from Guatemala, with a good number having very limited schooling. The transiency often results from one of the parents being deported. Then there's the question if the other parent will try to stay and make it work. Often, they'll end up moving back, because it's too expensive and hard with just one parent. Parents weren't bringing their kids to school, because they were afraid to go out. We are trying to connect our families with resources and information. We've brought in different foundations and support groups, as well as attorneys that work with immigration.

Each day, principals in EUSD address the same fundamental concerns faced by other school leaders across the country, vital concerns related to instructional effectiveness, equitable access to rigorous and relevant curriculum, the establishment of inclusive and engaging learning cultures, and the achievement of excellent learning results for all students. And they do so in ways that are revolutionizing student experience, student well-being, and the well-being of our planet. In Encinitas schools, learning tends to be integrated with nature, problem-/project-centered, appropriately individualized, and grounded in local places. Defining exactly what this looks like in practice is a deeply local affair.

Soon after becoming superintendent in 2009, former superintendent Dr. Timothy Baird formed a district Green Team comprised of interested parents, staff, and community members. Since that time, the Green Team and EUSD's environmental consultants (www.bckprograms.com/) have worked together to reduce the district's carbon footprint through facility upgrades, behavioral changes, and sustainability-focused educational programs. In fact, during the 2009–2010 school year, the district's Board of Trustees identified Environmental Stewardship as one of four key Pillars of Distinction that guide all district goals, along with Academic Excellence, Comprehensive Health and Wellness, and 21st-Century Learning.

A number of related Green Initiatives, depicted in the tree model below (See Figure 0.1) were funded, in part, through a $44 million bond extension passed in November 2010. Proceeds from the bond moneys have supported the provision of 21st-century classrooms (including infrastructure upgrades and technology tools for students and teachers), installation of solar panels and solar tubes, water reclamation systems for irrigation, replacement of inefficient heating and air condition systems, and water-saving upgrades to restroom fixtures at each school campus. The tree model depicts these and other initiatives that will be revisited in more detail throughout the chapters of this book. Dr. Baird shared the district's rationale for placing these goals and initiatives at the center of their work.

> You find so many pathways into learning from environmental stewardship. We're doing the right work, not just for our school district, but also for the world, and we're also finding amazing ways for kids to make real change and do real work. So, it is one of our four main pillars, central to the work that we do as a district.

Dr. Baird described the genesis of the district Green Initiative Model as a means to capture all aspects of whole school and district sustainability as currently practiced within the Encinitas Union School District.

> Before I arrived, they had amazing school gardens scattered around the district. In my first year, the Green Team came together and we started on garbage, but after we did garbage, we moved to lots of other things. We started to really look at air quality and energy. So, I said, "I need a graphic that pulls all our efforts together." [We] worked to put all these in a graphic that made sense [and depicted] all the things we're working on.

Of course, living systems models of schooling, like nature, are never static. EUSD leaders, teachers, students, parents, and community partners

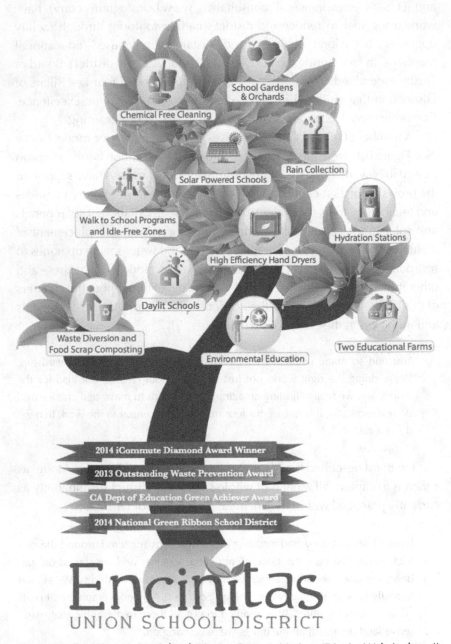

Figure 0.1 Encinitas Union School District Green Initiatives (District Website http://
www.eusd.net/green-initiatives/).

Chemical-Free Cleaning
The school district has invested in a chemical-free cleaning system that uses regular tap water and zaps it with an electrical charge to create liquid ozone, a powerful cleaning agent that kills germs as effectively as bleach and other chemical disinfectants without any of the harmful side effects. This green cleaning system eliminates the need for toxic cleaners, saves money and reduces waste.

Daylit Schools
Each school campus at EUSD now has daylighting devices (Solatubes) installed to allow natural light into the classrooms through protected tubes in the ceiling. So much light is allowed in using this method that there is often no need to turn on the electric lights. Research shows that use of these devices, called daylighting, increases students' overall test scores, improves moods and increases mental alertness. Daylighting also saves the school district money by reducing the energy demand of traditional lighting.

Educational Farms
In addition to the many garden spaces at each EUSD school, the district is also home to the nation's first certified organic school farm. EUSD's Farm Lab DREAMS Campus is an innovative indoor and outdoor educational campus for all students in the Encinitas Union School District. Farm Lab is the first in the nation to become a school-district-owned, certified-organic crop production farm supplying its own school lunch program. Ocean Knoll utilizes its one acre farm space to fortify the school's International Baccalaureate curriculum. Paul Ecke has transformed most corners of green space on the campus into farming experiments and hosts a weekend farmer's market every Sunday.

Environmental Education
The EUSD staff and its innovative teachers work alongside their environmental consultants at BCK Programs to offer numerous environmental education opportunities to the students at each EUSD school. Learning experiences in environmental stewardship include water and energy conservation, waste diversion, litter prevention, composting, environmental advocacy and the nationally recognized SWPPP Internship Program.

Figure 0.1 (Continued)

High Efficiency Hand Dryers

EUSD has installed high efficiency hand dryers in all student bathrooms, drastically reducing our consumption of paper towels on site. Replacing paper towels with hand dryers saves money and decreases our environmental impact by saving custodial labor time used to empty bathroom bins and unclog toilets, reducing waste going to the landfill and reducing the price of each hand dry to 1/20th the cost of using paper towels.

Hydration Stations

All EUSD schools have installed Hydration Stations for filling and refilling water bottles with filtered water. These stations encourage the use of refillable water bottles instead of single-use plastic bottles. By filling up reusable water bottles at our school's hydration stations we are not only giving our body the most perfect form of refreshment, we are also helping to reduce landfill waste, ocean pollution, and our carbon footprint.

Rain Collection

We have installed rainwater collection barrels at all of our school sites in an effort to conserve water and decrease runoff pollution. The collected water is stored and used as needed for irrigation in non-rain periods or as an educational tool for our school community. Rainwater is a renewable, sustainable, high quality water source. In addition, collecting water during storm events reduces flow speed which, in turn, reduces runoff into the storm drain keeping the ocean cleaner.

School Gardens and Orchards

All nine EUSD schools maintain at least one garden and several have orchards. These powerful environmental education tools provide a hands-on opportunity for students to learn the valuable skill of growing food, as well as fostering key values of teamwork and patience. Additionally, students engaged in growing edible plants are more willing to taste foods, exposing them to choices.

Solar Powered Schools

EUSD completed installing solar panels at all nine elementary schools in the summer of 2016. The panels are expected to cut the district's energy consumption by nearly 80% and save over $20 million dollars in future energy costs. In addition to students learning about power generated from a clean burning, abundant and renewable energy source like the sun, students are also exposed to the science and engineering behind the photovoltaic panels and inverter.

Figure 0.1 (Continued)

Waste Diversion and Food Scrap Composting
In order to reduce landfill waste, facilitate composting, and encourage lunchtime recycling, EUSD developed the SCRAP Cart (Separate, Compost, Reduce And Protect). The SCRAP Cart is used to teach students how to properly sort their lunchtime waste for composting, recycling and landfill. Since the introduction of the carts in 2012, lunchtime waste has been reduced at each school by over 80% saving the school district over $40,000 every year.

As part of the district's lunchtime waste management program, every EUSD campus is outfitted with large composting worm bins. Students learn the science of decomposition, while they divert food from their school's waste stream and turn it into a valuable garden amendment. Composting works in tandem with the school gardens reducing or even eliminating the need for fertilizer and reducing watering needs.

Walk to School Programs and Idle-Free Zones
We encourage walking and biking to school in our communities as an alternative to driving. Physical activity in the morning is known to improve academic performance, creativity and focus and to reduce student stress. In addition, incorporating physical activity into a child's daily routine is a good place to start fighting skyrocketing childhood obesity rates.

We protect the air around our schools by creating Idle-Free Zones at all of our schools to protect students from toxic pollutants released by cars idling near schools at drop off and pick up. Car idling contributes to health issues, contributes to smog and climate change, uses more gasoline than restarting your car, harms your engine and exhaust system, and increases vehicle maintenance costs.

Figure 0.1 (Continued)

constantly refine and extend their various initiatives; at the same time, they continuously scan the horizon, remaining open to the possibility of new partnerships and projects.

 Transforming Our Mental Models

Humanity currently faces a long list of profound challenges, including climate change, catastrophic weather events, devastating wildfires, biodiversity loss, population growth, an international pandemic, social inequities, and economic crises. Addressing these challenges requires we shift our worldview or mental models from one that sees humans as separate from nature to one that sees humans as integral with, and dependent upon, the natural world. Sustainability calls for this intentional and intelligent integration of human and ecological systems. As a field of study and a focal point for action, sustainability emphasizes the degree to which human beings choose to live within the ecological carrying capacity of planet Earth, presently and into the future (Bettencourt & Kaur, 2011). Sustainability scientists ask urgent questions:

> How do we survive and thrive on planet Earth?
> How do we take care of our planet, each other, and the resources we depend upon for our survival?
> How do we live responsibly so that those who come after us can live?

In the simplest terms, "Sustainability means making the world work. For everyone" (AtKisson, 2017, loc 121, Kindle).

To embrace the notion of schools as living systems, comprised of living beings who are deeply interdependent and embedded in local and global socio-ecological systems, educators must learn new habits of thought and practice (Kensler & Uline, 2017). The challenge, as we see it, is that mechanical systems and metaphors have influenced the design and management of our schools for more than a century (Senge, Cambron-McCabe, Lucas, Smith, & Dutton, 2012). For most of us, it's all we've ever known. Factory-like facilities, rigid schedules, bell notifications, defined grades, ability grouping, and unnecessary curricular narrowing in response to learning standards (for detailed discussion of alternatives to this unnecessary narrowing of curriculum, see Johnson, Uline, & Perez, 2019) remain pervasive across schools today, even as educators acknowledge the need for a fundamental redesign of the system in order to provide students 21st-century learning experiences.

The traditional factory model of education disassociated children and their learning from nature, each other, and their communities. Silos, metaphorically tall with very thick walls, still exist throughout school districts today. In too many schools, content exists in silos; teachers operate in silos; students learn in silos. Popular strategies associated with professional learning communities (PLCs) aim to deconstruct these silos, deprivatize education, and fuel learning (Stoll, Bolam, McMahon, Wallace, & Thomas, 2006). And yet, too often, these initiatives are implemented in name only, absent the necessary formal and frequent collaborative opportunities designed as a normal part of teachers' work days (Tichnor-Wagner, Harrison, & Cohen-Vogel, 2016).

Nature is characterized by interdependence and interconnectedness, not separation and isolation. Industrialized models of education tend to sever any real connection to nature and the outdoors by moving learning into rows of desks organized within four walls, sometimes even preventing outdoor distractions with windows covered or nonexistent. Understanding our schools as living systems results in two primary benefits (Kensler & Uline, 2017).

- First, we understand students, not as products of a 12-year assembly line, but as individual beings for whom the love of learning is an innate capacity. With this perspective, we stop demanding that learners perform as we direct. We stop blaming our students for disengaging. Rather, we realize that children and adults are voracious learners by nature and we design, manage, and lead for the conditions that allow this love of learning to flourish.

- Second, because our human communities are dependent on socio-ecological systems for life support, we accept responsibility for our actions. We realize that our daily actions contribute either to harming or enhancing the Earth's social communities and ecological systems and we consciously seek to minimize negative and maximize positive socio-ecological impacts.

School leaders have a critical role to play in developing more sustainable school practices, engaging students in the big questions and preparing them to discover and enact answers and solutions. As stated earlier, whole school sustainability, as a whole systems approach to K–12 education, integrates sustainability into all aspects of a school organization (Barr et al., 2014).

The actual number of schools who practice whole school sustainability remains a very small percentage of schools overall, comprising far less than 1% of K–12 schools. And yet, their trailblazing leaders are discovering all the ways whole school sustainability provides a high leverage strategy for addressing many 21st-century challenges, from student engagement and performance to climate change and community resilience.

The very goals of sustainability can "redefine the role of schools and their relationship with the community . . . [rendering] schools as a focal point where children, adults, and the community interact and learn together" (Henderson & Tilbury, 2004, p. 8). According to recent counts, communities across the globe benefit from 2,459 certified and 2,218 registered LEED (Leadership in Energy and Environmental Design) PK–12 school projects (US Green Building Council, 2020). Sustainability-related efforts are also reflected in the 4,300 National Wildlife Federation (NWF) Eco-Schools across the United States, serving 2.5 million students and over 116,000 educators, a growing cadre of Green School Alliance members (www.greenschoolsalliance.org/), and 595 U.S. Department of Education Green Ribbon School awardees, since 2012 (www2.ed.gov/programs/green-ribbon-schools/awards.html).

A recent study of living systems-minded school leaders found that 99% of respondents reported improvements in student engagement and 77% reported improvements in community engagement following their greening efforts (Sterrett, Imig, & Moore, 2014). These positive learning results, as well as dollar savings and environmental benefits, are increasingly attracting attention (Kensler & Uline, 2017). Schools, like the schools that comprise the Encinitas Union School District, become deeply rooted in their own place on the globe, ecologically and socially. They grow out of their unique context through the collective efforts of students, teachers, administrators, staff, parents, and community members. These sustainability-focused learning communities begin with a clear sense of purpose, growing deep and broad enough to inspire the long-term investment and commitment necessary to support such a fundamental shift in thinking and practice.

Keeping It Manageable

As readers consider this introduction, they may be moved to close this book, thinking, "Our school and district are so far from sustainable, I don't even know where to begin." Living systems-minded leadership does not take additional time; it simply requires recognizing and seizing opportunities for doing the work of school differently (Kensler & Uline, 2017; Uline & Kensler, 2019; Kensler & Uline, 2019). Dr. Baird offers practical advice to interested educational leaders:

> My answer now, when districts are asking about creating infrastructure [for whole systems approaches], is, 'Don't overthink it, just start it.' I started by

talking to different people at schools and saw that this was a passionate area for many people. It was a community value that was underdeveloped or under-realized.

In the spirit of Dr. Baird's advice, this book focuses on three straight-forward actions school leaders can begin to implement today. Aspiring and developing living systems-minded leaders can start in three manageable ways to move in the direction of whole school sustainability. They can begin to

- **Bring nature inside,**
- **Take students outside**, and
- **CARE**, by **C**ultivating and modeling **A**wareness, **R**esponsibility, and **E**mpathy.

In accordance with this Living Systems-Minded Leadership Model (Figure 0.2), our book is divided into three sections, corresponding to the actions presented above. Each section contains three associated strategies, described in detail and illustrated through specific practice-based

Figure 0.2 Living Systems-Minded Leadership Model.

examples, from Encinitas School District and other Living Systems-Minded Trailblazers throughout the United States.

Action 1: Bring Nature Inside

Nature, as we are using it here, is broadly defined as all the biological and physical elements of the world that are not human or created by humans. Humans experience nature either by spending time outside the built environment or by bringing nature into the built environment. Nature is increasingly incorporated into the built environment through expansive windows that flood the interiors with natural light and views of nature; and through including living plants, fish tanks, water fountains, etc. into working and learning spaces (Gillis & Gatersleben, 2015; Kellert, Heerwagen, & Mador, 2008). As we work to green existing school facilities and apply green principles to the design and construction of new schools, we articulate and advance sustainability goals and purposes. Likewise, when we utilize these sustainable schools as teaching tools, we extend our capacity to model socio-ecologically aware norms and practices (Taylor, 2009). Beyond conserving energy, decreasing stress on natural resources, preserving surrounding habitats, and reducing waste, we improve the ecological literacy of our students, teachers, administrators, and community members.

Strategy 1 LEAD: IMPLEMENTING NATURE-INSPIRED LEADERSHIP

Living systems-minded school leaders infuse their leadership with nature. They align their mental models, language, and behaviors with images of the natural world, rather than with industrial models of education. In fact, they intentionally identify and uproot persistent industrialized ways of thinking, speaking, and being, likely seeded early in their own educational experiences. Through their personal approaches to leadership, they expand their attention to school-wide programs and initiatives, aligning these efforts with principles of living systems.

Strategy 2 DESIGN: CHOOSING SUSTAINABLE BUILDING DESIGN ELEMENTS

Living systems-minded school leaders educate themselves about the possibilities for leveraging the physical learning environment on behalf of learning and teaching. Where resources become available for designing and constructing new facilities, leaders build a case for providing high-quality, sustainable school facilities. In situations where living systems-minded school leaders contend with older, existing facilities, they advocate for sustainability-focused renovations and retrofits when these become available to them. In all contexts, living-systems minded leaders assess their learning ecologies, taking particular and careful note of their buildings as critical both to occupant well-being and local and global environmental health.

Strategy 3 MAINTAIN & OPERATE: OPERATING AND MAINTAINING HEALTHY, SAFE, SUSTAINABLE LEARNING ENVIRONMENTS

Living systems-minded leaders, together with their facilities colleagues, manage healthy, safe, and sustainable learning environments in ways that reduce energy, conserve natural resources, and minimize waste. To the greatest degree possible, they implement green operation and maintenance routines, and seek out opportunities for leveraging the facility as a three-dimensional textbook.

Action 2: Take Students Outside

Strategies for reconnecting students with nature might begin simply with opening the doors to the outside and reintroducing recess. Emerging research suggests improved student behavior, learning focus, and academic performance follow daily recess, unstructured play in the outdoors (Bauml, Patton, & Rhea, 2020). Beyond recess, academic learning can also occur productively while deeply embedded in the outdoors. Students learn basic content, in addition to gaining deep insight into how the world works as an integrated, interdependent whole. Numerous recent reviews of research demonstrate that contact with nature is associated with overall health and well-being, including specific aspects of emotional, physical, social, and cognitive well-being (Hartig, Mitchell, de Vries, & Frumkin, 2014; Kuo, 2015; Louv, 2008; Russell et al., 2013). The evidence is substantial—children benefit from contact with nature.

Strategy 4 TEACH: PREPARING TEACHERS TO TEACH IN NATURE

Living systems-minded school leaders provide ongoing, job-embedded opportunities for teachers to learn the requisite knowledge and skills for facilitating student learning in nature. They challenge teachers to move outside their classroom comfort zone and provide the necessary resources and supports to ensure teachers' success in doing so.

Strategy 5 LEARN: INVITING STUDENTS AND TEACHERS TO LEARN IN NATURE

Living systems-minded school leaders intentionally disrupt the traditional architecture of instruction. They open doors and invite learning to deliberately spill out beyond school walls. Living systems-minded school leaders create space and time for teachers, parents, and community members to experience the ways nature is associated with overall health and well-being, including specific aspects of emotional, physical, social, and cognitive well-being, all of which are foundational to students' engagement in learning. The stark realities of the COVID-19 pandemic have underscored the benefits of learning outside in nature. In late summer 2020, as local districts developed their plans to reopen schools, Dr. Anthony Fauci (head the National Institute of Allergy and Infectious Diseases) recommended "Get[ting] as much outdoors as you can" (https://abcnews.go.com/Health/wireStory/fauci-schools-outdoors-72359724). Further, time in nature also presents valuable opportunities for students to learn complex concepts and develop important academic skills (Camassoa & Jagannathan, 2018).

Strategy 6 PLAY: RESTORING NATURE PLAY INTO THE SCHOOL DAY

Living systems-minded school leaders value play as a pathway for learning. They prioritize recess and encourage teachers to take their students outside for unstructured learning time. They ensure play spaces include a rich variety of features, inviting exploration, challenge, creativity, and restoration. Living systems-minded school leaders cultivate a school culture that honors individual needs and empowers teachers and students to make appropriate choices for their own and collective well-being. They know outdoor, unstructured play provides opportunities for learning critical motor skills and social skills while also restoring attention and energy for learning.

Action 3: Care

In their own particular place on Earth, living systems-oriented school leaders hold themselves to account as their schools' lead learners. They take time to clarify their own sense of purpose as educators and challenge themselves to investigate the implications of current societal and environmental challenges for their work as 21st-century school leaders. This reflection prompts them to consider their role as civic leaders across the social, economic, *and* ecological systems within which their schools are nested. The potential scope of their responsibilities can seem overwhelming, and so, they engage others in crafting a laser-focused vision for their work, a vision grounded in intimate knowledge of, and *CARE* for, the place their students call home. In this way, they are able to **C**ultivate and model **A**wareness, **R**esponsibility, and **E**mpathy. When people are able to connect their daily work to meaningful, purposeful aims, motivation soars. They feel passionate about their contribution to their vision for a healthy, flourishing community.

Strategy 7 MODEL: CULTIVATING AND MODELING AWARENESS, RESPONSIBILITY, AND EMPATHY

Living systems-minded school leaders gain intimate knowledge of, and model, *CARE* for the place their students call home. They cultivate and model awareness, responsibility, and empathy throughout their entire school community. Living systems-minded school leaders *cultivate* the learning capacity of their members, including teachers, students, parents, and the community at large. They lead all members in developing a deep *awareness* of, and sense of *responsibility* for, their unique social and ecological context. They develop *empathy* for communities upstream and downstream and for other human and nonhuman inhabitants across the planet.

Strategy 8 PARTNER: BUILDING CARING PARTNERSHIPS

Living systems-minded school leaders cultivate caring partnerships that have potential to revitalize their communities. Embracing whole school sustainability, as means to maximize student learning, also encourages a sense of responsibility for the well-being of the community-at-large and the natural world upon which it depends. As school leaders pursue the mutually reinforcing aims of maximizing learning and developing community, they discover powerful partners in reimagining day-to-day school life and in securing future life on Earth.

Strategy 9 START SMALL: STARTING SMALL AND STAYING ANCHORED IN A VISION OF VIBRANT, FLOURISHING, SUSTAINABLE SCHOOLS

Living systems-minded school leaders facilitate the development of shared visions for sustainability. These visions are expansive, trans-formative, and motivating. Through articulation and implementa-tion of such visions, living systems-mined leaders reveal, direct, and strengthen the interdependent connections between each action they and members take to realize their vision of vibrant, flourishing, sus-tainable schools.

At the conclusion of each strategy, we present readers with an associated *Leadership Design Challenge* as inspiration to begin, or expand, their own green school efforts. These practical and attainable projects provide school leaders opportunities to, in the words of Encinitas Union School District former Superintendent Dr. Timothy Baird, "[not] overthink it, just start it." In addition, readers will have opportunity to learn about actions taken by *Living Systems-Minded Trailblazers* from across the United States who, like the leaders, teachers, students, and community members in the Encinitas School District, are partnering with nature in numerous ways to design, lead, and manage vibrant, flourishing, sustainable learning communities.

 ## Conclusion

Through deep investigation of the possibilities contained within three straight-forward actions, as revealed through one particular school district's story in one place on planet Earth, as well as through the stories of other Living Systems-Minded Trailblazers, we hope to help school leaders see how pur-suing these actions can transform their schools and school districts into vibrant centers of learning and socio-ecological responsibility. We think readers will see that these mutually reinforcing aims provide powerful leverage for school improvement, as well as for improvement of life on Earth.

References

AtKisson, A. (2017). *Sustainability is for everyone*. Hofheim: Center for Sustainability Transformation.

Barr, S. K., Cross, J. E., & Dunbar, B. H. (2014). *The whole-school sustainability framework: Guiding principles for integrating sustainability into all aspects of a school organization*. Washington, DC. Retrieved from http://centerforgreenschools.org/sites/default/files/resource-files/Whole-School_Sustainability_Framework.pdf

Bauml, M., Patton, M. M., & Rhea, D. (2020). A qualitative study of teachers' perceptions of increased recess time on teaching, learning, and behavior. *Journal of Research in Childhood Education*, 1–15. doi: 10.1080/02568543.2020.1718808

Bettencourt, L. M., & Kaur, J. (2011). Evolution and structure of sustainability science. *Proceedings of the National Academy of Sciences of the United States of America*, *108*(49), 19540–19545. doi:10.1073/pnas.1102712108

Camassoa, M. J., & Jagannathan, R. (2018). Improving academic outcomes in poor urban schools through nature-based learning. *Cambridge Journal of Education*, *48*(2), 263–277.

Gillis, K., & Gatersleben, B. (2015). A review of psychological literature on the health and wellbeing benefits of biophilic design. *Buildings*, *5*(3), 948–963. doi:10.3390/buildings5030948

Hartig, T., Mitchell, R., de Vries, S., & Frumkin, H. (2014). Nature and health. *Annual Review of Public Health*, *35*, 207–228. doi:10.1146/annurev-publhealth-032013-182443

Henderson, K., & Tilbury, D. (2004). *Whole-school approaches to sustainability: An international review of whole-school sustainability programs*. Canberra, Australia: Australian Research Institute in Education for Sustainability.

Johnson, J. F., Uline, C. L., & Perez, L. (2019). *Teaching practices from America's best urban schools: A guide for school and classroom leaders* (2nd ed.). New York: Routledge/Taylor and Francis Group.

Kellert, S. R., Heerwagen, J. H., & Mador, M. L. (2008). *Biophilic design: The theory, science, and practice of bringing buildings to life*. Hoboken, NJ: Wiley.

Kensler, L. A. W., & Uline, C. L. (2017). *Leadership for green schools: Sustainability for our children, our communities, and our planet*. New York: Routledge/Taylor and Francis Group.

Kensler, L. A. W., & Uline, C. L. (2019). Educational restoration: A foundational model inspired by ecological restoration. *International Journal of Educational Management, 33*(6), 1198–1218.

Kuo, M. (2015). How might contact with nature promote human health? Promising mechanisms and a possible central pathway. *Frontiers in Psychology, 6*, 1093. doi:10.3389/fpsyg.2015.01093

Louv, R. (2008). *Last child in the woods: Saving our children from nature deficit disorder*. Chapel Hill, NC: Algonquin Books of Chapel Hill.

Russell, R., Guerry, A. D., Balvanera, P., Gould, R. K., Basurto, X., Chan, K. M. A., & Tam, J. (2013). Humans and nature: How knowing and experiencing nature affect well-being. *Annual Review of Environment and Resources, 38*(1), 473–502. doi:10.1146/annurevenviron-012312-110838

Senge, P. M., Cambron-McCabe, N., Lucas, T., Smith, B., & Dutton, J. (2012). *Schools that learn (updated and revised): A fifth discipline fieldbook for educators, parents, and everyone who cares about education*. New York: Crown Business.

Sterrett, W. L., Imig, S., & Moore, D. (2014). U.S. Department of Education Green Ribbon Schools: Leadership insights and implications. *Journal of Organizational Learning and Leadership, 12*(2), 2–18.

Stoll, L., Bolam, R., McMahon, A., Wallace, M., & Thomas, S. (2006). Professional learning communities: A review of the literature. *Journal of Educational Change, 7*(4), 221–258.

Taylor, A. (2009). *Linking architecture and education: Sustainable design of learning environments*. Albuquerque: University of New Mexico Press.

Tichnor-Wagner, A., Harrison, C., & Cohen-Vogel, L. (2016). Cultures of learning in effective high schools. *Educational Administration Quarterly, 52*(4), 602–642. doi:10.1177/0013161x16644957

Uline, C. L., & Kensler, L. A. W. (2019). Whole district transformation: Leading systems change for sustainability. In C. Schechter, H. Shaked, & A. Daly (Eds.), *Leading holistically: How schools, districts and states improve systemically*. New York: Routledge/Taylor and Francis Group.

US Green Building Council. (2020). Personal communication.

ACTION

1

Bring Nature Inside

Nature, as we are using it here, is broadly defined as all the biological and physical elements of the world that are not human or created by humans. Humans experience nature either by spending time outside the built environment or by bringing nature into the built environment. Nature is increasingly incorporated into the built environment through expansive windows that flood the interiors with natural light and views of nature and through including living plants, fish tanks, water fountains, etc. into working and learning spaces. As we work to green existing school facilities and apply green principles to the design and construction of new schools, we articulate and advance sustainability goals and purposes. Likewise, when we utilize these sustainable schools as teaching tools, we extend our capacity to model socio-ecologically aware norms and practices. Beyond conserving energy, decreasing stress on natural resources, preserving surrounding habitats, and reducing waste, we improve the ecological literacy of our students, teachers, administrators, and community members.

Lead

Cultivate Living Systems-Minded Leadership

Most of us have grown up experiencing organizational structures and routines designed according to industrial models of operation. Our schooling is no exception. Industrial models have long been applied to schools and school districts, resulting in a mismatch between structure and function. Schools and school districts seek to advance learning (a living system) as a primary aim; at the same time, they treat learning more like a mechanical system (Senge, Cambron-McCabe, Lucas, Smith, & Dutton, 2012). This fundamental contradiction results in undesirable and unacceptable outcomes when it comes to student learning, graduation rates, teacher retention rates, and more. Adults and children may withdraw, disengage, and even leave school altogether, to avoid the dysfunctional structures and routines that, too often, deny them any sense of individual agency (see Kensler & Uline (2017, 2019) for more in depth discussion). As Murphy noted, "Schooling for students is profoundly voluntary. Children have to 'go to school.' The decision to 'do schooling' is substantially their own" (2015, p. 725).

Mental models, conceptions or images of how the world works, develop throughout our lifetimes. Frequently operating unconsciously, these mental models guide our thoughts, judgments, and behaviors. Shifting one's mental models, and eventually one's leadership, from ineffective industrial models to more effective living systems models, requires intentional effort that counters forces and trends toward less connection with nature (Soga & Gaston, 2016). Living systems-minded school leaders uproot unconscious, industrial models of learning and leading and replace them with more vibrant, life-sustaining models. This strategy outlines various steps living systems-minded leaders take as they actually bring nature into their leadership practice as they (1) connect with nature, (2) examine their mental models, and (3) apply lessons from nature to their work.

Connect With Nature

Our use of the term *nature* aligns with a general definition of nature or the natural world, as that which is not specifically human or built from non-living material by humans. Of course, humans design, build, and manage living landscapes through activities such as gardening, farming, and restoration of degraded ecosystems. We include these natural spaces, influenced by human activity, in our conception of nature. Nature's ecosystems illustrate an awe-inspiring capacity for vitality, creativity, adaptation, and resilience—all properties we seek to cultivate in our schools. An Encinitas Union School District (EUSD) principal described how she learned this fundamental truth about the natural world when attending a recent conference on green schools.

> The presenter reminded us that humans have been on the Earth for such a short period of time. She explained that the Earth really knows, on its own, how to take care of itself. If we listen to what nature does to solve problems, there lies our answer.

If leaders wish to transform schools into more vibrant places of learning, they can learn much by connecting with nature themselves, welcoming nature into their development as healthy humans and leaders. In this section, first, we summarize the benefits enjoyed by individuals who connect with, and spend time in, nature. Next, we note opportunities for learning that emerge from this greater awareness of, and closer relationship with, nature.

Well-Being

School leaders experience significant levels of stress in their work, a trend that appears to be increasing (Wang, Pollock, & Hauseman, 2018). The daily demands of the job necessitate outlets for developing resilience and managing stress (Carpenter, 2020). Nature offers educational leaders a readily available antidote. Researchers report a long list of associated benefits for individuals when they connect with, and spend time in, nature, including many attributes of health and well-being, as well as cognitive ability and pro-environmental behaviors (Capaldi, Passmore, Nisbet, Zelenski, & Dopko, 2015; Keniger, Gaston, Irvine, & Fuller, 2013; Louv, 2008, 2011). Capaldi et al. (2015) noted the mutually reinforcing relationship between spending time in nature, referred to as nature contact, and nature connectedness, "a subjective sense of connection with the natural world" (p. 2). Spending time in nature fuels our connectedness with nature, at the same time a deep connection with nature correlates with spending more time in nature. Research also demonstrates benefits associated with children spending time in and connecting with nature. We will explore these benefits further in Strategies 5 and 6. For now, we remain focused on adults, particularly educational leaders.

Although extended periods of time in wilderness settings offer potential for meaningful personal growth and team building (Superville, 2019), connecting with nature can also happen during brief moments each and every day. Passmore and Holder (2016, p. 543) found that simply "paying increased attention to everyday nature significantly increased individual well-being," regardless of whether participants reported having a prior sense of connectedness with nature. They further explained that these positive effects did not require spending more time in nature, participants "simply noticed, and attended to, the nature they encountered in their daily routines" (p. 543). Thus, countering stress and cultivating resilience does not have to be a costly endeavor. Educational leaders can simply develop habits of noticing.

As we consider potential strategies to encourage school leaders' self-care, and minimize their risk of burnout, connecting with nature presents a cost-effective approach, both in terms of time and money. The associated outcomes, from spending even a few minutes immersed in nature, are powerful. Connor Moriarty, founder of Reset Outdoors, presented the THRIVE model in a TEDx talk. He captured the multifaceted outcomes associated with spending time in, and connecting with, nature (Moriarty, 2020). We THRIVE with time spent outdoors and connecting with nature.

T—Thoughts. Just looking out a window at the landscape or even simply images of nature in one's office can restore one's attention and focus. Taking a walk in the park is even better (Berman, Jonides, & Kaplan, 2008; Stevenson, Schilhab, & Bentsen, 2018).

H—Health. Time in nature is associated with a broad range of health and well-being benefits (Capaldi et al., 2015; Kuo, 2015)

R—Resilience. One's ability to bounce back after physical, emotional, mental challenges is improved with time spent in nature (Capaldi et al., 2015; Ingulli & Lindbloom, 2013)

I—Interdependence. Interdependence speaks to our social connections, as well as our connection with nature. John Donne's familiar line, "No man is an island, entire of itself; every man is a piece of the Continent, a part of the main", reminds us of our interdependencies. Research tells us that our perceptions of social cohesion, social interdependence, are positively associated with time spent in nature (Shanahan et al., 2016).

V—Vitality. Vitality is beyond just being healthy. It is that feeling you have when you're on top of the world, when you know you can face any challenge in your path. We increase our vigor and vitality with time spent in nature (Hyvonen et al., 2018; Korpela, De Bloom, Sianoja, Pasanen, & Kinnunen, 2017).

E—Empathy. Empathy, the capacity to feel with others and consider their perspectives, plays a critical role in leadership,

---

organizations, and our local and global communities. In order for society to meet the challenges of climate change, members must develop empathy for nature. Research suggests that spending time in nature, with fellow humans, is associated with increased empathy for them and also for nature (Brown et al., 2019; Zhang, Piff, Iyer, Koleva, & Keltner, 2014).

Learning

Rampant industrialization, with little attention to the needs and properties of living systems, impedes the natural operation of ecological principles and degrades ecosystem functioning. When individuals spend more time in nature, they learn more about these natural systems upon which we all depend. They begin seeing more examples of ecological principles in action. Ecological principles govern healthy living systems and, since human social systems are living systems, these principles apply to the work of leading and managing learning organizations, specifically schools. The ecological principles we described in *Leadership for Green Schools* (Kensler & Uline, 2017), inspired by Capra (2002) and the Center for Ecoliteracy, are by no means the only principles governing healthy systems, but they are a great starting place for our observation and learning. We have limited space to discuss them here, but have done so in depth previously (Kensler, 2012; Kensler & Uline, 2017). Attentive observation and creative reflection will reveal opportunities for applying these principles to the work of transitioning from industrial models of schooling to living systems models of schooling. We may actually see opportunities for transforming our own beliefs, or mental models, about how schools ought to function.

Living systems are nested systems

Networks, partnerships, and diversity support resilience

Self-organization and creativity emerge from within the system itself

Matter cycles and energy flows through living systems

Feedback sustains dynamic balance

Examine Mental Models

Scharmer (2018) quoted a prominent CEO's insight about organizational change, "The success of an intervention depends on the interior condition of the intervener" (p. 7). Our mental models of how the world works, operating on both conscious and unconscious levels, characterize this interior condition. In the case of educational leaders as "interveners", mental models influence how they lead changes in school structures, how they facilitate their school's learning culture, how they design and manage their school facilities, and how they collaborate with community partners. Table 1.1 compares common mental models associated with mechanical and living systems approaches of education. One's individual experiences, learning within schools characterized by mechanical systems, make it likely that they may embrace mental models across both categories. In the absence of careful scrutiny, mental models are unlikely to change. It takes intentional reflection and critical examination to replace mechanical systems mental models with living systems mental models. As beliefs about leading change, learning, facilities, and community evolve, living systems-minded leaders become ready to apply lessons learned from nature.

 ## Apply Lessons From Nature

When noticing and connecting with nature, educators glean insights that inform leadership development and practice. In other words, lessons from nature have direct application to the work of leadership (Covey, Merrill, & Jones, 1998; DeLuca, 2016). Biomimicry is the practice of applying solutions found in nature to a vast array of problems across the human experience, including production of materials, design of organizational structures and processes, and, yes, even implementation of leadership practices (Benyus, 1997; DeLuca, 2016). Covey et al. (1998) quoted William Wordsworth, "Come forth into the light of things, Let Nature be your teacher" (pp. 10–11). Dr. Baird described how lessons from nature informed leadership decisions within EUSD.

> Perspective is important. We started looking at biomimicry and how we build this notion of imitating nature into the work that we're doing. What an amazing way to start thinking about things, because nature has a lot of different solutions to problems that we're out looking for solutions for.

Table 1.1 Comparison of Underlying Assumptions/Mental Models for Mechanical and Living Systems Approaches to Education.

		Mental Models Underlying Mechanical Systems Approach	Mental Models Underlying Living Systems Approach
Leading Change		- Bureaucratic structures guide and control change - External visions inspire change - External mandates and accountability systems direct and monitor change - Answers to complex problems lie outside schools and districts with experts and policy makers	- Living systems change from within - Vision emerges as an internal property of the system through engaging processes - Networked, flexible, responsive structures facilitate change - Each school and district transforms from within, supported with capacity building strategies rather than mandated solutions
Learning	Teachers and Administration	- Adults are already experts; adult learning is a lesser concern than student learning	- Continuous adult learning is facilitated, nurtured, supported
	Students	- Children are deficient and need fixing - Children are passive learners - Learning is best disconnected from nature and the real world	- Children are whole and capable learners with agency - Children actively engage in their own learning - Learning is integrated with nature and the real world
	Curriculum	- Teachers deliver unnecessarily narrowed, standardized content via rigid pacing guides - Curriculum exists in silos and is teacher/classroom centered	- Teachers utilize guiding standards to design culturally, socially, and personally responsive; place-, problem-, project-based learning opportunities - Curriculum is integrated across content areas and grade levels
Facilities	3-D Textbook	- Facilities house students with little to no curricular connections	- Building/grounds are integral to curriculum and teaching/learning programs

(Continued)

Table 1.1 (Continued)

		Mental Models Underlying Mechanical Systems Approach	Mental Models Underlying Living Systems Approach
	Building	- Traditional building functions with little to no concern for health, well-being, environmental responsibility	- Green/Living building characteristics serve health, well-being, and environmental responsibility
	Grounds	- Mono-cultured lawns and ornamental plants are managed with fertilizer, herbicides, and pesticide	- Gardens, landscapes, and native habitats focus on ecological restoration and food production
Community	**Conception of Community**	- Schools are closed systems - School are self-sufficient and professionalized	- Schools are open systems - Schools embrace the complexities of social, cultural, and ecological interdependence
	Engagement with Community	- Students learn about public issues without engaging with these real issues - Students learn little about their local context - Teachers and students are buffered from external disruptions - Some parents participate in schools via traditional PTAs, parent nights, etc., while many others are not included	- Teachers guide student learning about, and engagement with, real public issues at all grade levels, activating student voice and agency - Students become experts on their local communities - Parents are welcomed and active partners in their child's learning and growth
	Responsibilities to Community	- Schools aim to independently deliver a uniform product at the completion of each grade and school level - Schools graduate future workers - Schools maintain student order and discipline	- Schools engage in ongoing co-creation with their communities - Schools play an active role in community and ecological restoration - Order and discipline are a natural outcome of students' choosing to engage in challenging and relevant learning

Applying lessons from nature aids in cultivating vibrant school communities where learning thrives. Across the nine strategies presented in our book, we offer many examples of lessons from nature that apply to leading schools. We begin with three fundamental nature-informed lessons for consideration—(1) restore rather than reform; (2) navigate rather than control; and (3) engage in systems thinking. This is by no means an exhaustive list. In fact, the lessons we might garner from nature are bounded only by the limits of our curiosity, openness, and willingness to consider their appropriate application to our work.

Restoring Rather Than Reforming

As our mental models shift towards living systems-minded alternatives, we discover whole new ways of thinking about, and addressing, the shortcomings and dysfunctions inherent within our current educational systems. Ecological restoration, an applied field within ecology, teaches that degraded ecosystems can be restored through careful redesign that facilitates natural functioning. As just one example, a dried-out meadow that was once an ecologically valuable wetland can return to a flourishing wetland if we dismantle the artificial barriers (dikes, culverts, dams, etc.) that block the natural water flow. Hawken (2007) explained ecological restoration as "extraordinarily simple: You remove whatever prevents the system from healing itself" (p. 189). Kensler and Uline (2019) presented a model of educational restoration—the practice of redesigning educational systems such that barriers to children's natural love of learning are removed. Educational restoration is not another approach to educational reform because, rather than tweaking the industrial model (characteristic of most educational reform approaches), educational restoration removes fundamental barriers left behind by industrialized approaches to managing learning (see Table 1.1).

EUSD's living systems-minded leaders continuously challenge themselves to remove barriers that impede the natural functioning of their learning ecosystem. Consider their recent efforts to renew the district's commitment to equity and excellence. District leaders abandoned bureaucratically structured decision rules in favor of a more horizontal, interdependent approach, seeking active presence and focused engagement on the part of all members. An Equity Team, comprised of representatives from the district's nine schools, as well as principals from the districts' three Title 1 schools, came together to draft a three-year plan, aimed at ensuring every student in EUSD has opportunity to learn deeply and perform to their highest potential. In response to a district-wide invitation, a group of 50 leaders, teachers, staff, and parents joined the Equity Team through a series of ongoing meetings that began in the spring of 2020 (via ZOOM during the COVID-19-related shut down). This group of 50 has provided input that will be used in assessing needs and creating district-wide learning opportunities to guide the work. At the close of each gathering, the group identified action items to be implemented prior to the next meeting. Assistant Superintendent for Educational Services, Dr. Amy Illingworth described the natural flow of energy that resulted from an invitation to communicate openly and honestly about difficult topics.

> The group of 50 was really passionate about changing everything right away. We had to remind them that this is slow work. It's hard, and sometimes it doesn't feel like you get very far, very quickly. So, we are making sure that there's some sort of immediate action they can take, whether it's personally, in their classroom, in their grade level, or at the school level.

Industrial models of reform have relied substantially on top-down directives, with expectations for rapid organizational change. However, rarely have these approaches resulted in deep, transformative change. Restoring community connections and engaging healthy processes will take more time, but the desired changes are more likely to take root and flourish. Living systems-minded leaders embrace the promise of going slow to go fast and understand that if they do not have the time to go slow, they will very likely have to make the time to try again. Restoring the innate learning capacity of our organizations will make them more responsive, adaptive, and resilient; they will be more capable of meeting a never-ending flow of challenges.

Navigating Rather Than Controlling

Living systems are complex, dynamic systems. Learning is a property of living systems and, thus, functions as a complex and dynamic system, in and of itself. Attempts at managing and controlling learning, as so many educational reforms and programs aim to do, typically fail (Murphy, 2017; Payne, 2008). Instead, educators are better positioned to navigate, rather than control, the complex, dynamic systems that make up their school communities. "The key to dealing with change is to have a changeless core" (Covey et al., 1998, p. 73). A 'changeless core' speaks to nature's laws and/or principles, "If you think in terms of principles, are true to principles, and exercise faith in the results, they will eventually come to pass" (p. 74). Covey et al.'s book is a compilation of individual reflections about life and leadership shared by participants from a nature-based leadership program. With time on a river, participants shared how their fundamental understanding of "go with the flow" evolved from a sense of fluid passivity, just going along for a ride, to a deeper appreciation for the importance of active presence, focused engagement, and working with continuously changing conditions, rather than attempting to ignore, or even control, change. On the river, each boater needs the autonomy to partner with the river. In a similar way, each school leader needs the autonomy to partner with their educational ecosystem In Figure 1.1, a student waters the learning garden, an important part of her school's educational ecosystem.

Assistant Superintendent Illingworth described what happened when district leaders relinquished control and embraced their principals as rightly autonomous leaders of their schools' learning ecosystems.

> Each school has their own unique identity and site-specific focus area, with alignment to our district pillars. For instance, one of our schools is an International Baccalaureate (IB) school. We also have two Dual Language Immersion schools. Our schools have their own personalities, with learning personalized for students, based on the school's identity. These identities serve as brands, which help our community members determine the best learning environment for their children.

Dr. Baird emphasized the way in which district leaders support and navigate individual principals' best judgments, providing necessary flexibility and reaping the subsequent benefits.

> This is not the district waving the wand and saying, "You're this school or you're that school", because you can't do that. It has to come from who the school is. It has to be part of what their nature is, what they want to focus on, and what engages them. As leaders, we ask, "How do we tap into those passions?

Figure 1.1 Gardens throughout EUSD nurture students' love of learning.

Where do we find the phenomena that drive this work?" And, some schools have changed over time. They are a different school than they were five years ago. I love that, because they learn from each other as they change.

A principal described her leadership experience in EUSD.

Here they (district leaders) really believe in the teachers, and they believe in the principals. You have a lot autonomy to take your program and make it fit with the culture of your building. Teachers can then take an idea and work together to put it into practice.

With autonomy, principals and teachers have the opportunity to fully engage with their learning ecosystems; they are free to shift from fluid passivity to active presence, focused engagement, and working with con-tinuously changing conditions.

Engaging in Systems Thinking

Nature teaches us that everything is connected to everything else. Our interconnected, interdependent socio-ecological systems are complex and dynamic. They are continuously in flux. If we treat them as if they are predictable and controllable machines, then we will be forever surprised and frustrated by unexpected and undesirable outcomes. Systems thinking teaches us how to engage more productively with complex, dynamic systems (Fullan, 2005; Senge et al., 2012).

> The discipline of systems thinking provides a different way of looking at problems and goals—not as isolated events but as components of larger but less visible structures that affect each other. To understand a system is to understand those interrelationships and how they recur and change over time. A school district is a system with many interrelated components: everything from the design of the buildings to the habits and attitudes of the people who work there to the policies and procedures imposed by the state and the com-munity, as well as such implacable forces as available money and student population growth or decline. When you see how these affect each other, you can act far more effectively.
>
> (Senge et al., 2012, p. 124)

 Conclusion

Spending time in, and connecting with, the natural world teaches us how we might bring nature into our leadership development and practice. As we do this, our well-being and resilience improve, and we gather myriad lessons from nature that change our thinking about ourselves and others, our organizations, our communities, and our leadership. "We need to re-align our hearts and souls, our practices and culture, ourselves and our organizations with Nature" (DeLuca, 2016, p. 83).

Covey et al. (1998) underscored the power of choice; individuals have choices about what to notice, what opportunities to engage, and how to lead. As Dr. Baird suggested, nature teaches us how to view our organizations through different perspectives, and once armed with these different perspectives, we are able to discover new questions and novel solutions. Nature inspires us to think differently about how we lead. Day and Leithwood (2007), in their reflections on decades of studying school leadership, explained,

> . . . we want to dig a bit deeper below the surface of our empirical evidence than do these explanations. Such digging has led us to the view that our empirical evidence about the nature of successful principal leadership can be partly explained by the basic metaphors leaders and their colleagues hold about their organizations. . . . Two competing metaphors—"organizations as machines" and "organizations as living systems"—are features in Wheatley's[1] explanation for both organizations and leadership that differ radically in their functioning and outcomes. *The work of our successful principals strongly suggests that they thought of their organizations as living systems, not machines.*
>
> (p. 200, emphasis added)

 Leadership Design Challenges

1. **Connect with nature**—Challenge yourself to notice nature (a tree, the sky, the color of the grass, etc.) as you move throughout your day. Document your noticing by taking a picture with your phone. At the end of the day, take a few minutes to reflect on your day's pictures— What did you notice? What connections might you make to your leadership?

2. **Examine your mental models**—Challenge yourself to notice and reflect on the mental models that inform your leadership. What words do you repeatedly use when discussing leadership issues and/or leading? What images of leadership do these words evoke in you? In others? Use Table 1.1 to help you get started.

3. **Apply lessons from nature to your leadership practice**—Challenge yourself to select one or two powerful insights per week that you gained from (#1) connecting with nature and (#2) examining your mental models to weed out industrial modes of thinking, learning, and/ or communicating. Over each week's time, intentionally and consistently seek to root out one or two industrial approaches to leading and replace them with more living systems-minded approaches to leading.

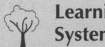

Learning From Living Systems-Minded Trailblazers

Missouri Outdoor Leadership Experience

The Leadership Academy, part of the Missouri Leadership Development System within the Missouri Department of Elementary and Secondary Education, focuses on developing effective school leaders who:

- Engage with school and community
- Serve as catalysts for meaningful and productive change
- Recognize and develop excellent instruction
- Create and sustain a culture of continuous learning (Leadership Academy, 2020)

The 36th class of the Leadership Academy will take place over the 2020/2021 school year. Each Leadership Academy class has had the opportunity to participate in a three-day Outdoor Leadership Experience. This innovative programming takes place at a YMCA Camp adjacent to the Mark Twain National Forest. Participants are housed in rustic, climate-controlled cabins and many of the leadership development activities take place outside. Activities include a variety of team challenges, individual reflections, and group debriefings.

We interviewed three graduates of the program who, at the time, were serving as leaders in Missouri public schools, including John Fortney, Jennifer Hayes, and Megan Stryjewski. They all spoke passionately about their time in the Outdoor Leadership Experience and conveyed the dramatic impact it has had on their leadership. Being in nature was an essential component of the experience. For many participants, spending this much time in nature may be brand new and offers a way to connect with themselves and others differently than in their typical day-to-day. Megan explained,

> I do believe that being outdoors, closer to nature and away from the rest of the world, helps you to reflect. So, having that opportunity to go

sit by the lake or walk in the woods, where you don't have great cell reception, helps people to reflect and maybe connect differently. . . . And then they make this connection with what we ask teachers and what we ask students to do. All the time [we ask them] to put their trust and their faith in us and do something that they thought they would never want to do. And so, immersing ourselves in these experiences and understanding the feelings and reactions we might have, opens our eyes as leaders and helps us to maybe have some empathy and understanding for our teachers and students, as well as just carry out initiatives in a different way.

Jennifer explained that the Outdoor Leadership Experience was deeply transformative. She returned "renewed, refreshed, excited about the things that I'd done, excited about my new network of people and just really energized to continue figuring out how to be a good leader." She continued,

I am 100% more open to hearing everyone's ideas. I am much slower to make decisions. I've made peace with the fact that I have to say, "I'm not sure. Give me some time. I need to gather some information before we decide what we're going to do." I have learned that the way to build success is to build a strong foundation and go slowly and make sure you have consensus.

John rediscovered his love of the outdoors and the value of spending time in nature as means to managing stress and restoring his capacity to lead. He described hiking nearly every weekend following the Outdoor Leadership Experience.

. . . just trying to escape to find those moments of clarity, because, for me, that was what OLE helped me learn again—how to center myself. So, regardless of the day I'm having, I put a pack on, I grab my dog, and get on a trail. I don't know if I'm subconsciously solving problems, but I'm more centered and more focused and ready to deal with anything when I get off the trail and back to my day job.

The Missouri Outdoor Leadership Experience invites school leaders to discover new lessons about themselves as leaders while immersed in nature. Nature teaches them to welcome the unexpected, pause, reflect, and then act. These and many more lessons help them cultivate healthier and happier school cultures where learning is more likely to thrive.

Note

1. Margaret Wheatley is an author who contrasted organizations as machines with organizations as living systems in her book, *Leadership and the New Science*, in addition to her other works listed on www.margaretwheatley.com

References

Benyus, J. M. (1997). *Biomimicry*. New York: William Morrow.

Berman, M. G., Jonides, J., & Kaplan, S. (2008). The cognitive benefits of interacting with nature. *Psychological Science, 19*(12), 1207–1212.

Brown, K., Adger, W. N., Devine-Wright, P., Anderies, J. M., Barr, S., Bousquet, F., . . . Quinn, T. (2019). Empathy, place and identity interactions for sustainability. *Global Environmental Change, 56*, 11–17. doi:10.1016/j.gloenvcha.2019.03.003

Capaldi, C. A., Passmore, H. A., Nisbet, E. K., Zelenski, J. M., & Dopko, R. L. (2015). Flourishing in nature: A review of the benefits of connecting with nature and its application as a wellbeing intervention. *International Journal of Wellbeing, 5*(4), 1–16. doi:10.5502/ijw.v5i4.1

Capra, F. (2002). *Hidden connections*. New York: Doubleday.

Carpenter, B. (2020). Reframing self-care as altruistic: An interruption of self-denial. *UCEA Review*, Summer, 1–4.

Covey, S. M. R., Merrill, A. R., & Jones, D. (1998). *The nature of leadership*. Salt Lake City, UT: Franklin Covey Company.

Day, C., & Leithwood, K. (Eds.). (2007). *Successful principal leadership in times of change: An international perspective*. Dordrecht, The Netherlands: Springer.

DeLuca, D. (2016). *[Re]aligning with nature: Ecological thinking for radical transformation*. Ashland, OR: White Cloud Press.

Fullan, M. (2005). *Leadership and sustainability: Systems thinkers in action*. Thousand Oaks, CA: Corwin.

Hawken, P. (2007). *Blessed unrest: How the largest movement in the world came into being and why no one saw it coming*. New York: Penguin Group.

Hyvonen, K., Tornroos, K., Salonen, K., Korpela, K., Feldt, T., & Kinnunen, U. (2018). Profiles of nature exposure and outdoor activities associated with occupational well-being among employees. *Frontiers in Psychology, 9*, 754. doi:10.3389/fpsyg.2018.00754

Ingulli, K., & Lindbloom, G. (2013). Connection to nature and psychological resilience. *Ecopsychology, 5*(1), 52–55. doi:10.1089/eco.2012.0042

Keniger, L. E., Gaston, K. J., Irvine, K. N., & Fuller, R. A. (2013). What are the benefits of interacting with nature? *International Journal of Environmental Research and Public Health, 10*(3), 913–935. http://dx.doi.org/10.3390/ijerph10030913

Kensler, L. A. W. (2012). Ecology, democracy, and green schools: An integrated framework. *Journal of School Leadership, 22*(4), 789–814.

Kensler, L. A. W., & Uline, C. L. (2017). *Leadership for green schools: Sustainability for our children, our communities, and our planet*. New York: Routledge/Taylor & Francis Group.

Kensler, L. A. W., & Uline, C. L. (2019). Educational restoration: A foundational model inspired by ecological restoration. *International Journal of Educational Management, 33*(6), 1198–1218. doi:10.1108/ijem-03-2018-0095

Korpela, K., De Bloom, J., Sianoja, M., Pasanen, T., & Kinnunen, U. (2017). Nature at home and at work: Naturally good? Links between window views, indoor plants, outdoor activities and employee well-being over one year. *Landscape and Urban Planning, 160*, 38–47. doi:10.1016/j.landurbplan.2016.12.005

Kuo, M. (2015). How might contact with nature promote human health? Promising mechanisms and a possible central pathway. *Frontiers in Psychology, 6*, 1093. doi:10.3389/fpsyg.2015.01093

Leadership Academy. (2020). *Flyer*. Retrieved from https://dese.mo.gov/sites/default/files/Leadership-Academy.pdf

Louv, R. (2008). *Last child in the woods: Saving our children from nature deficit disorder*. Chapel Hill, NC: Algonquin Books of Chapel Hill.

Louv, R. (2011). *The nature principle: Human restoration and the end of nature-deficit disorder*. Chapel Hill, NC: Algonquin Books.

Moriarty, C. (2020, May). What Kant missed about human nature: And finding it in nature. *TEDx Lehigh River Conferences*. Retrieved from www.youtube.com/watch?v=JB1bq8X9paI

Murphy, B. G. (2017). *Inside our schools: Teachers on the failure and future of education reform*. Cambridge, MA: Harvard Education Press.

Murphy, J. (2015). The empirical and moral foundations of the ISLLC standards. *Journal of Educational Administration, 53*(6), 718–734. doi:10.1108/jea-08-2014-0103

Passmore, H.-A., & Holder, M. D. (2016). Noticing nature: Individual and social benefits of a two-week intervention. *The Journal of Positive Psychology, 12*(6), 537–546. doi:10.1080/17439760.2016.1221126

Payne, C. M. (2008). *So much reform, so little change: The persistence of failure in urban schools*. Cambridge, MA: Harvard Education Press.

Scharmer, C. O. (2018). *The essentials of theory u: Core principles and applications*. Oakland, CA: Berrett-Koehler.

Senge, P. M., Cambron-McCabe, N., Lucas, T., Smith, B., & Dutton, J. (2012). *Schools that learn (updated and revised): A fifth discipline fieldbook for educators, parents, and everyone who cares about education*. New York: Crown Business.

Shanahan, D. F., Bush, R., Gaston, K. J., Lin, B. B., Dean, J., Barber, E., & Fuller, R. A. (2016). Health benefits from nature experiences depend on dose. *Scientific Reports, 6*, 28551. doi:10.1038/srep28551

Soga, M., & Gaston, K. J. (2016). Extinction of experience: The loss of human-nature interactions. *Frontiers in Ecology and the Environment, 14*(2), 94–101. doi:10.1002/fee.1225

Stevenson, M. P., Schilhab, T., & Bentsen, P. (2018). Attention restoration theory II: A systematic review to clarify attention processes affected by exposure to natural environments. *Journal of Toxicology and Environmental Health, Part B: Critical Reviews, 21*(4), 227–268. doi:10.1080/10937404.2018.1505571

Superville, D. R. (2019, October 29). What principals learn from roughing it in the woods. *Ed Week*. Retrieved from www.edweek.org/ew/articles/2019/10/30/what-principals-learn-from-roughing-it-in.html

Wang, F., Pollock, K. E., & Hauseman, C. (2018). School principals' job satisfaction: The effects of work intensification. *Canadian Journal of Educational Administration and Policy*, *185*, 73–90.

Zhang, J. W., Piff, P. K., Iyer, R., Koleva, S., & Keltner, D. (2014). An occasion for unselfing: Beautiful nature leads to prosociality. *Journal of Environmental Psychology*, *37*, 61–72. doi:10.1016/j.jenvp.2013.11.00

Design

Choose Sustainable Building Systems and Design Features

Living systems-minded school leaders educate themselves about the possibilities for utilizing physical learning environments to improve teaching and learning, as well as occupant well-being. They act strategically to maximize these, too often missed, opportunities. Where resources become available for designing and constructing new facilities, leaders build a case for providing high-quality, sustainable school facilities as a means for realizing children's current learning potential and for advancing their future quality of life. In cases where living systems-minded school leaders contend with older, existing facilities, they seek ways to bring nature inside by uncovering and augmenting sources of natural light, creating indoor gardens and scenes of nature, highlighting views to the outdoors where they exist, and exploring ways to increase the availability of fresh odor-free air. They also advocate for sustainability-focused renovations and retrofits when these are deemed necessary. In the face of COVID-19 pandemic-related challenges, it is worth mentioning that these sustainable design choices increase the capacity of school facilities and systems "to absorb, respond, and adapt to disruptive change . . . design[ing] to make the world better in the long-term but prepar[ing] to absorb continued disturbance in the short-term" (Williams, 2020, p. 39). In all these contexts, living systems-minded leaders assess their learning ecologies, taking particular and careful note of their school buildings as critical, both to occupant well-being and local and global environmental health.

This strategy outlines various steps living systems-minded leaders take as they seek to envision and enact sustainable building systems and design features throughout their school facilities. Living systems-minded leaders embrace sustainable design principles as they (1) understand what makes a green building green; (2) learn how and why to bring nature inside; (3) reuse, recycle, and refurbish existing facilities where possible; and (4) harness the knowledge and energy that results when they connect the facilities and the curricular/instructional sides of the academic house.

Understand What Makes Green School Buildings Green

Widely accepted definitions describe green schools as healthy and high performing (Collaborative for High Performance Schools (CHPS), n.d.). Green school buildings, whether built new or made more sustainable through targeted improvements, create healthy environments that are conducive to learning while saving energy, resources, and money (US Green Building Council's Center for Green Schools, n.d.). These facilities apply sustainable energy conservation techniques and tools across building systems. They are also designed and/or upgraded with particular attention to ambient features such as acoustic, thermal, and visual comfort.

According to the National Research Council (2007), green schools have two complementary goals and resulting outcomes. First, green schools aim to "support the health and development (physical, social, intellectual) of students, teachers and staff by providing a healthy, safe, comfortable, and functional physical environment" and, second, green schools seek to have "positive environmental and community attributes"

(p. 2). When approached in an integrated fashion, the design of a green school, including the nature of school site; the building orientation; the building envelope (exterior building shell); the heating, ventilation, and air conditioning systems; the acoustics; and the lighting systems, result in a physical learning environment with appropriate levels of moisture, ventilation, air quality, noise, and lighting (LPA Architects, 2009; National Research Council, 2007). In the best cases, these green buildings function as interactive tools for learning, three-dimensional textbooks, if you will (Kong, Rao, Abdul-Rahman, & Wang, 2014; Nair & Fielding, 2005; Taylor, 1993).

In November 2010, voters in the Encinitas Union School District passed a $44 million bond extension aimed at renovating and upgrading existing facilities throughout the district. At the time, the school district's existing school facilities ranged in age from 15 to 50 years, placing necessary building modernizations (roof repairs, ADA compliance measures, classroom improvements) first on the list of needs. Second on the list of needs were additional technology tools for students and teachers. Dr. Baird explained,

> We were kind of in the dark ages related to our technology use. We needed to create a technological infrastructure and, also, although not through the bond dollars, we needed to prepare our staff to take care of, help expand, and best utilize that technology infrastructure.

Green upgrades, including the development of their Farm Lab (see Figure 0.1. Encinitas Union School District Green Initiatives, beginning on p. 22), placed number three and four on the list of facilities' needs. The facility initiative sought to increase energy efficiency and conserve natural resources across all schools by providing solar panels as a renewable energy source, installing solar tubes for lighting, replacing inefficient heating and air condition systems, utilizing catchment systems to reclaim water for irrigation, and upgrading restroom fixtures (See Figure 0.1. Encinitas Union School District Green Initiatives, beginning on p. 22). These forward-thinking, green improvements demonstrate that new, state-of-the-art facilities are not a requirement for going green in schools. Rather, a shift in mindset on the part of EUSD leadership, from "take-make-waste" to "refuse-reduce-reuse-recycle", prioritized a sustainability-focused investment of their precious capital funds. Dr. Baird noted, "All of our schools have solar. We now have solar tubes at every school." Julie Burton, Coordinator of Innovation and Farm Lab Development at EUSD, shared, "We've opened up the classroom walls [at the Farm Lab] by putting in barn doors so that you get an indoor or outdoor feel. It's another way to bring in more natural light and fresh air." And one EUSD principal underscored the point, "I'm sure you saw the solar panels over our parking lot. So again, that's how sustainability is lived out in our district, through that [ever-present] green focus." Figure 2.1 shows one of the water catchment systems also present on school campuses across the district.

Where new construction is possible, living systems-minded school leaders are increasingly making facilities-related choices that will improve occupant well-being. In fact, a recent study demonstrated better occupant cognitive function in certified green, over non-green, buildings (Allen et al., 2016). The certified green buildings in this study were built to strict environmental and energy standards using the U.S. Green Building Council's LEED (Leadership in Energy and Environmental Design) rating system, thus reducing their impact on the environment while also improving conditions for human well-being and performance to thrive. Although we do not yet have comparably rigorous studies in schools, we know that teaching and learning depend on healthy cognitive function. The implications for our school facilities, and the students and teachers who spend the majority of their waking hours there, are powerful.

Bring Nature Inside the School Building

Green schools maximize the benefits of connecting with nature by bringing nature's life-giving properties into school building design. The emerging field of biomimicry informs the work. As defined in Strategy 1, biomimicry is the practice of applying solutions found in nature to a vast array of problems across human experience, including production of materials and structures, as well as organizational processes (Benyus, 1997; DeLuca, 2016). Fresh odor-free air, abundant natural light, inspiring views of nature, and indoor gardens are just some examples of biomimicry-inspired design features common in green schools.

Delivering an Abundance of Clean Indoor Air

Indoor air quality, which results from the presence/absence of outdoor and indoor air pollutants, thermal comfort, and sensory loads (odors,

Figure 2.1 Water catchment systems reclaim water for irrigation.

freshness), can affect the heath of children and adults and may affect student learning and teacher productivity (National Research Council, 2007, p. 54). Pollutants and allergens in indoor air, including mold, dust, bacterial and fungal products, volatile organic compounds (VOCs), and particulate matter, have been associated with asthma and other respiratory symptoms, as well as eye, nose, and throat irritations; headaches; and fatigue (National Research Council, 2007). Elevated CO_2 levels have also been associated with student absenteeism (Shendell et al., 2004) and reduced rates of attention in primary school students (Coley & Greeves, 2004).

Even before school construction is completed and the building is occupied, dust and outgassing contaminants, introduced by conventional construction materials, influence IAQ in negative ways. Sustainably sourcing naturally made building materials, including flooring, acoustic ceiling tiles, insulation, signage, and wall paneling, eliminates these contaminants.

Providing Ample Access to Daylight and High-Quality Artificial Lighting

Throughout a school day, students engage in a wide variety of learning experiences and tasks, most of which include some visual dimension. In fact, "[t]he visual qualities of a learning environment are some of the most crucial building aspects to design properly since children depend heavily on sight in the learning process" (Baker & Bernstein, 2012, p. 10). Both the quantity and quality of light are important factors in design decisions, with wide agreement on appropriate quantity firmly established and additional research required on issues regarding quality (Baker & Bernstein, 2012).

High-quality natural light helps to create a sense of physical and mental comfort with benefits beyond aiding sight (Barrett & Zhang, 2009). Building orientation constitutes a fundamental design choice for the control of light, given the variations in available daylight at east-, west-, north-, and south-facing facades. Additional daylight design strategies, including the size and placement of windows in combination with clerestories, skylights, solar tubes, and light shelves, optimize daylight distribution and bring light deeper into a given space. Frosted glass and easily operable shades help to eliminate glare and provide teachers necessary control of lighting across the day and for different learning activities.

Windows and views, as well as various forms of light, have been linked to student health, learning, and behavior, with daylight offering a more positive effect on student outcomes, potentially due to its biological effects on the human body. For example, a study by The Heschong Mahone Group (1999) analyzed data from 2,000 classrooms across school districts in Orange County, California; Seattle, Washington; and Fort Collins, Colorado. Test results of 21,000 students, demographic data, architectural plans, aerial photographs, the presence of skylights, maintenance records, and daylighting conditions were among the factors considered. Findings revealed students in daylit classrooms showed greater improvements across one school year in math and reading than did students in window-less classrooms.

Ensuring High Quality Classroom Acoustics

Acoustic performance involves the control of ambient and external noise within an enclosed space in order to provide high-quality conditions for producing and receiving desired sounds. The control of noise and the quality of auditory perception primarily determine the nature and quality of the acoustic environment (Barrett & Zhang, 2009). A substantial body of research underscores the importance of minimizing background noise and maximizing speech intelligibility in order to ensure students have optimal opportunities to learn (Berg, Blair, & Benson, 1996; Maxwell & Evans, 2000; Knecht, Nelson, Whitelaw, & Feth, 2002). Too often, students and teachers struggle within poor acoustically performing classrooms (Feth & Whitelaw, 1999; Sato & Bradley, 2008). School design decisions that take account of location to avoid excessive external noise, consider layout of academic spaces to minimize intrusive internal noise, and plan for room features that incorporate sound absorbent materials to reduce sound reverberations ensure high-quality acoustic performance.

Affording Thermal Comfort Within the Classroom

Thermal comfort within the built environment is maintained through control of temperature, humidity, and airflow. A well-designed building envelope helps to mediate external conditions when combined with high-efficiency

49

and double-paned windows, siting for solar access and ventilation, and high R-value (thermal resistance) insulation (Taylor, 2009). High-performance heating, ventilation, and air conditioning (HVAC) systems help to minimize distractions and allow for focused learning. When a classroom is too hot, cold, or damp, students and teachers have difficulty settling in. Understandably, they'd rather be somewhere else, and so, instead of fully engaging in the work at hand, their thoughts stray to some more inviting and comfortable place (Kensler & Uline, 2017).

Incorporating "Inside/Outside" Design Features

In a way, living systems-minded school leaders turn their school campuses inside/out and outside/in, blurring the lines between built and natural environments. In so doing, they invite students to observe how these spaces are necessarily connected and interdependent, opening the door (literally and figuratively) for nature-based learning to take place. Dr. Baird provided examples of the ways EUSD connected built and natural learning environments throughout the district.

> So, I think it's really an inside/outside thing, because we have nature all over our campuses. Our gardens function as outdoor classrooms. Teachers will often take students to an outdoor classroom for a given lesson, for example, into a harvesting garden. They will also bring food from that garden inside to the nutritional lab and continue the lesson, as students prepare and cook the food. We have these kitchens on several of our campuses. One of our principals is engaged in a renovation of their library, working with her staff to open it up to the outside. Learning will flow from a maker space, to a coding area, to a green room for video. Then, they're opening it up to an outdoor garden with the installation of sliding doors. So, it really is indoor/outdoor, similar to the barn doors at Farm Lab, where kids move from inside to outside and back. It just blurs that line (See Figure 2.2).

A recent review of the psychological literature related to well-being and biophilic design found growing evidence for occupants' "reduced stress, improved productivity, and improved well-being" (p. 957) in buildings that "incorporate natural features and systems into the built environment in order to provide human beings with their much-needed exposure to nature" (Gillis & Gatersleben, 2015, p. 949).

Figure 2.2 Barn doors connect indoor and outdoor learning at Farm Lab.

Reuse and Adapt Existing Facilities

Reusing and adapting existing facilities reduces landfill waste and avoids the use of new natural resources, with the associated fiscal investment and energy costs (Filardo, 2016, p. 6). EUSD was not only creative in reusing and adapting existing structures, but they also utilized these up-cycling practices as learning opportunities for students. Dr. Baird provided an example:

> At Farm Lab we added two new classrooms. We took portables that were headed for the landfill and put a challenge out to all of our upper-grade students. Flora Vista kids were the ones that signed up for it first and said, "We want to do this!" Their teachers were committed to it. I gave them an actual budget and I said, "You need to design these classrooms as research and design spaces. Come up with how you're going to outfit them, what the decking space will look like in between them. Remember we have to stay green and, by the way, you have to stay under budget."

Dr. Baird indicated that the "classrooms became textbooks." Under the direction of an area architect, the students developed blueprints. They prepared Excel spreadsheets. They called manufacturers. They negotiated prices. He explained,

> In addition to working with area architects, the students video-conferenced with Maker space designers in Washington, D.C. The project extended over

three months, with the students applying math skills, writing skills, research skills, and presentation skills. It's what learning should look like. We developed a two-minute "This is what I learned" video clip, and it's everything we would want to hear as an educator. "I learned that I have to work with other people to get the best product. I learned that real work isn't easy and sometimes you fail and you learn from that. I learned that I've got to look at multiple sources before I decide. You don't just go to Amazon to buy stuff."

Connect the Two Sides of the Academic House

School districts are complex bureaucracies with traditional hierarchies and typical silos. Quite relevant to sustainability-related reforms is the typical division between the curriculum and facilities departments. Many green school initiatives address building improvements and sit squarely in the facilities department. However, green building projects have the potential to serve as powerful learning opportunities for students at all levels. Dr. Baird spoke pointedly about the potential for missed opportunities when the academic house remains divided.

> The fact that facilities people don't talk to educational people and the educational people don't talk to the facilities people is a crime. One of our best educators is Gerry Devitt (Director of Facilities, Maintenance, Operations and Grounds). He loves to work with kids, and his guys are great teachers for the students. The more you blend facilities with the learning experience, the richer the experience is.

Living systems-minded school leaders build bridges between the historically divided curriculum and facilities sides of the academic house. When the facilities team and the curriculum team work in concert, they inform and enrich each other's work, helping occupants live more consciously and responsibly in their buildings. To the degree educators collaborate with their facilities colleagues, they extend the reach of these functions in ways that "help learners clarify their own roles and responsibilities with respect to sustainability-related issues and develop efficacy and agency to act on these responsibilities" (Nolet, 2016, p. 74).

When EUSD embarked on their green planning process, all relevant players were invited to participate, including the facilities director, food services personal, the assistant superintendent for curriculum, IT, the district's green consultants, building principals, and teachers. Dr. Baird acknowledged that students were a missing element, and that "they should have been there." Over time, students have played primary planning roles, with the classroom design project mentioned above as a case in point. Baird shared, "Students have not been shy about bringing new ideas to us, to the School Board, to City Council." We provide examples of students' meaningful and authentic work and stewardship across the nine strategies presented in this book.

Build a Compelling Case for Going Green

Understandably, planning a sustainable school "raises the stakes," with design, construction, and management practices that are new to school districts (Gelfand, 2010, p. 19). Skeptics continue to maintain that green systems and technologies are too complicated for school maintenance staff; green materials aren't durable; and, most frequently, that the cost of going green is too high. Living systems-minded leaders are wise to prepare themselves to respond with evidence of the benefits, both in terms of their students' and teachers' well-being and performance, and in terms of their bottom lines.

Consider a study of primary schools in Great Britain. Edwards (2006) compared 42 pairs of schools, green versus conventional, with similar characteristics in terms of geography, size, grade level, SES, and numbers of students with special needs and for whom English is a second language. The study considered five sets of performance factors, including performance on standardized tests, absenteeism, bullying, teacher turnover and teaching days lost due to illness, and qualitative interview data from teachers and department heads regarding their work experiences. The schools designated as green were designed based on ecological principles, demonstrating evidence of energy efficiency, health (both physically and psychologically), comfort, responsiveness, and flexibility. Research findings demonstrated a 3–5%improvement in test scores consistently displayed in all but one green school, with more significant improvements for younger students who stay in one classroom longer. Data also revealed lower teacher absenteeism and turnover rates, reduced rates of student absenteeism and bullying in green schools, as well as an enhanced image on the part of the community for their school. Holistic green design strategies, where design, construction, and curriculum practices were integrated, appeared to offer greater advantages over concentration on a single aspect of green design. In addition, schools that prioritized daylight and natural ventilation generally outperformed other schools, both in urban and rural contexts (Edwards, 2006).

In terms of upfront investment, good daylighting, ventilation, site design, and effective integration of building systems should not cost more (LPA Architects, 2009). Where certain materials, controls, and sensors are expensive by themselves, these components should be carefully considered within the context of the entire system (Gelfand, 2010). A national report, authored by Gregory Katz in 2006, documented the financial costs and benefits of green schools compared to conventional schools. Data gathered between 2001 and 2006 from 30 green schools built in 10 states revealed the cost per square foot for green design was $4 more than conventional design, while the operating benefits of green design realized approximately $68 per square foot in savings. Typically, green schools cost 1% to 2% more than conventional buildings, with an average cost premium of 1.7% or $3 more per square foot (Kats, 2006). This cost differential is often referred to as the "green premium", resulting from the increased cost of sustainably sourced materials, more efficient mechanical systems, and other high-performance building features (Kats, 2006). Any increase in upfront costs raises concerns for already financially strapped districts, and yet, growing evidence suggests, in the case of green schools, each dollar spent today realized two dollars in future savings (Gelfand, 2010). Sustainability-focused organizational practices require we adopt the routine of life cycle costing, factoring in the costs of installation, operation, and disposal over the entire life of a material or system. Rather than relying upon first-cost analyzes, this more thorough price accounting reveals the actual cost of green versus convention school buildings (Eley, 2006; Gelfand, 2010), helping living systems-minded school leaders build the case for future returns on initial investment.

Green schools realize an average of 33% energy reduction over conventionally designed schools as a result of efficient lighting, greater use of daylighting and sensors, more efficient heating and cooling systems, and better insulated walls and roofs (Kats, 2006). They experience a 32% reduction in water use as a result of efficient plumbing fixtures, green roofs, water catchment systems, and the like (Kats, 2006). Durable construction materials and thorough building commissioning processes also reduce operations and maintenance costs. In fact, a study of the costs and benefits of greening California public buildings across 40 state agencies found an $8 per square foot savings in operations and maintenance costs over a twenty-year period (Kats, Alevantis, Berman, Mills, & Perlman, 2003). Green school leaders should avail themselves of these compelling data; at same time they amass their own evidence of building performance over time.

In a recent article for *School Business Affairs*, Tim Cole, Sustainability Manager for the Virginia Beach City School District (VBCPS), tells the story of how VBCPS "moved toward an educational and operational model that embraces the triple bottom line of social, economic, and environmental effects when measuring profit and loss" (Cole, 2015, p. 32). Cole reported on the resulting savings realized by the school district.

> VBCPS has been moving in a direction that eased the transition to tough economic times. . . . By constructing [eight] new buildings according to LEED criteria and focusing on Energy Star and performance contract work for HVAC and lighting in existing buildings, VBCPS has reduced energy costs. Since 2006, the school division has increased in square footage by 9%—to 10.6 million square feet. At the same time, energy use per square foot has decreased by 21%.
>
> (Cole, 2015, p. 33)

VBCPS's experiences also challenge the accuracy of the green premium. When compared with construction costs across the states of Delaware, the District of Columbia, Maryland, Virginia, and West Virginia, all eight VBCPS LEED projects were built for less than the regional average. Cole is quick to point out that most of the comparison schools are non-LEED buildings (Cole, 2015).

Green project certification, resulting from a third-party verification process, helps to make a compelling case for going green. Certification systems, such as USGBC's Leadership in Energy and Environmental Design (LEED) and Collaborative for High Performance Schools (CHPS) provide public confirmation of sustainability efforts, building public trust in the quality of the outcomes. In addition, a comprehensive building commissioning process (a systematic quality assurance procedure that begins with the first stages of planning and extends through design, construction, and post-occupancy) increases the chances that the finished product will success-fully meet the needs of students, their teachers, their community, and the natural world. In addition, Educational Commissioning informs teachers, students, even parents and community partners, of the design intentions, helping them to best leverage all aspects of the physical learning environment (Lackney, 2005). We will revisit the commissioning process in greater detail under Strategy 4.

When EUSD leaders launched efforts to secure taxpayer support for the 2010 bond, they brought together a team of parents, staff members, and community partners who helped spread a consistent message about the need for facility improvements. Approximately 63% of voters approved the ballot measure, with green upgrades key among their reasons for voting yes. Dr. Baird explained how the school district's decision to go green ultimately made their case.

> I think one of the things that really compelled a lot of people *were* the green aspects. Now, the way you present green aspects can be through a financial lens. So, we talked about how spending bond dollars to put solar on all of our campuses was going to save so many dollars every operational year. This is money we were now spending on an electric bill. We couched it in two different ways. One, we're going to do the right thing for the environment, and, two, we're going to save money for the district in the long run. People were pretty happy about that!

Conclusion

According to Taylor (2009), "The ideal educational environment is a carefully designed physical location composed of natural, built, and cultural parts that work together to accommodate active learning across body, mind, and spirit" (Taylor, 2009, p. 31). When schools, as significant places in students' lives, reflect sustainable design principles, they become rich, varied, and dynamic learning environments within which students gain a sense of agency as local and global citizens who can make a difference for their own and each other's futures. Further, these green schools are "ideal place[s] to plant an idea that is meant to propagate throughout a community" (Gelfand, 2010, p. 7). And so, students, through their physical engagement with the environmental features of their school—vegetable gardens, demonstration kitchens, compost piles, energy system monitors, ponds—model responsibility, stewardship, and active engagement in learning.

They derive meaning and purpose from these carefully conceived spaces and gain confidence in teaching others about the important discoveries they make.

 Leadership Design Challenges

1. **Schedule a meet and greet with facilities colleagues**.
   Design and build an open door between the two sides of the academic house. Seek regular counsel from members of your school district's Facilities, Maintenance, and Operations Department. Go to the experts and ask them how existing design features and building systems are maintaining the health of your school buildings, and decreasing your schools' carbon footprints, or not. Ask them how you can best help to support their work in improving and maintaining healthy school campuses. Listen. Inviting these experts to the table builds capacity for informed decisions, as well as organizational support for future actions.

2. **Design an indoor/outdoor learning space on your school campus**.
   Create an outdoor classroom. You might choose a weather station, a solar and wind energy station, a nature trail, a greenhouse, a water harvesting systems, or perhaps merely a circle of tree stumps placed in a beautiful spot on your campus. Link this outdoor classroom to your interior learning environment by placing it where it is visible through classroom windows, accessible via exterior doorways, and/or conceptually and instructionally aligned to science or nutrition labs. Establish a schedule whereby each class has the opportunity to utilize this indoor/outdoor classroom space at least once each week. See Green Schoolyards America for resources related to research, policy, and support (www.greenschoolyards.org).

3. **Choose one biomimicry-inspired design feature common in green schools. Build a problem-based unit through which students build a case for the importance of this feature to their learning**.
   Students focus on a specific feature they discover to be insufficient within their learning environment, such as fresh odor-free air, abundant natural light, inspiring views of nature, indoor gardens, ergonomic furniture, etc. Research and design processes drive their investigation, collection of relevant data, and conversations with school district and

community experts. Resources and/or materials, obtained through business partners or existing bond funds, might make implementation of their proposals possible.

 ## Learning From Living Systems-Minded Trailblazers

Sacramento Unified School District's Project Green Initiative engages student-led Green Teams in conducting "green" audits of their school facilities. With the help of teachers, parents, district facilities staff, and community experts, the teams drafted recommendations for green improvements, ranging from replacing outdated windows to installing low-flow plumbing fixtures. One SUSD high school principal spoke about the importance of inviting district operations and maintenance departments to engage with students in their assessments of building conditions.

> The operations and maintenance folks were so excited to be a part of the process. Some of these gentlemen have been here for forty years. They're craftsmen. We have nine different shops lead by nine different foremen who possess specialty trades. This represents an enormous store of knowledge to bring to the table, and nobody had ever asked.

These place- and problem-based projects engage student voice, as teams analyze how much energy they consume, how much water they use, the kinds of materials their buildings are made of, how they handle waste, what kinds of cleaning and chemicals they use, and present their recommendations to a panel of local experts from the fields of architecture, engineering, energy, land-use planning, and water management. The panel judges the exhibits, presentations, and written papers using a rigorous scoring rubric. In its first year, fifteen schools were awarded funds for these green facilities projects.

The team who scored highest for its presentation and report advanced strong rationale for providing the school with automatic hand dryers, rain barrels, solar tubes, and upgrades to irrigation and HVAC systems. The judging panel voted to allocate up to $550,000 for this one school's projects. In all, the first 15 award-winning presentations were allocated $5 million in capital improvement funds (made available through existing bond monies). Over the first three years of implementation, 31 separate Project Green projects were completed, 13 projects were underway, and an additional 20 projects were forthcoming, with Project Green continuing today.

References

Allen, J., MacNaughton, P., Satish, U., Santanam, S., Vallarino, J., & Spengler, J. (2016). Associations of cognitive function scores with Carbon Dioxide, ventilation, and volatile organic compound exposures in office workers: A controlled exposure study of green and conventional office environments. *Environmental Health Perspectives, 124*(6), 805.

Baker, L., & Bernstein, H. (2012). *The impact of school buildings on student health and performance*. Washington, DC: McGraw-Hill Research Foundation and The Center for Green Schools.

Barrett, P., & Zhang, Y. (2009). *Optimal learning spaces: Design implications for primary schools*. SCRI Report No.2, SCRI, Salford.

Benyus, J. M. (1997). *Biomimicry*. New York, NY: William Morrow.

Berg, F. S., Blair, J. C., & Benson, P. V. (1996). Classroom acoustics: The problem, impact, and solution. *Language, Speech, and Hearing Services in Schools, 27*, 16–20. doi:10.1044/0161-1461.2701.16

Cole, T. (2015). Why sustainability makes good economic sense. *School Business Affairs*, 32–34.

Coley, P. A., & Greeves, R. (2004). *The effects of low ventilation rates on the cognitive function of a primary school class*. Report R102 for DfES. Exeter University.

Collaborative for High Performance Schools (CHPS). (n.d.). Retrieved from https://chps.net/dev/Drupal/node

DeLuca, D. (2016). *[Re]aligning with nature: Ecological thinking for radical transformation*. Ashland, OR: White Cloud Press.

Edwards, B. W. (2006). Environmental design and educational performance. *Research in Education, 76*, 14–32.

Eley, C. (2006). High performance school buildings. In H. Frumkin, R. J. Geller, & I. L. Rubin (Eds.), *Safe and healthy school environments* (pp. 341–350). New York: Oxford University Press.

Feth, L., & Whitelaw, G. (1999). *Many classrooms have bad acoustics that inhibit learning*. Columbus, OH: Ohio State.

Filardo, M. (2016). *State of our schools: America's K-12 facilities 2016*. Washington, DC: 21st Century School Fund.

Gelfand, L. (2010). *Sustainable school architecture*. Hoboken, NJ: John Wiley & Sons, Inc.

Gillis, K., & Gatersleben, B. (2015). A review of psychological literature on the health and wellbeing benefits of biophilic design. *Buildings, 5*(3), 948–963. doi:10.3390/buildings5030948

Heschong Mahone Group. (1999). *Daylighting in schools: An investigation into the relationship between daylighting and human performance*. Fair Oaks, CA: Heschong Mahone Group.

Kats, G. (2006). *Greening America's schools: Costs and benefits*. Retrieved from www.usgbc.org/showfile.aspx?DocumentID-2908

Kats, G., Alevantis, L., Berman, A., Mills, E., & Perlman, J. (2003). *The costs and financial benefits of green buildings*. Sacramento, CA: A Report to California's Sustainable Building Task Force.

Kensler, L. A. W., & Uline, C. L. (2017). *Leadership for green schools: Sustainability for our children, our communities, and our planet*. New York: Routledge/Taylor and Francis Group.

Knecht, H. A., Nelson, P. B., Whitelaw, G. M., & Feth, L. L. (2002). Background noise levels and reverberation times in unoccupied classrooms predictions and measurements. *American Journal of Audiology, 11*, 65–71.

Kong, S., Rao, S., Abdul-Rahman, H., & Wang, C. (2014). School as 3-D textbook for environmental education: Design model transforming physical environment to knowledge transmission instrument. *Asia-Pacific Education Researcher, 23*(1), 1–15. doi:10.1007/s40299-013-0064-2

Lackney, J. (2005). Educating educators to optimize their school facility for teaching and learning. *Design Share*. Retrieved from www.designshare.com/index.php/articles/educational-commissioning/

LPA Architects. (2009). *Green school primer*. Victoria, Australia: The Images Publishing Group.

Maxwell, L. E., & Evans, G. W. (2000). The effects of noise on pre-school children's pre-reading skills. *Journal of Environmental Psychology, 20,* 91–97.

Nair, P., & Fielding, R. (2005). *The language of school design: Design patterns for 21st century schools.* Minneapolis, MN: Designshare, Inc.

National Research Council. (2007). *Green schools: Attributes for health and learning.* Washington, DC: The National Academies Press.

Nolet, V. (2016). *Educating for sustainability.* New York: Routledge.

Sato, H., & Bradley, J. S. (2008). Evaluation of acoustical conditions for speech communication in working elementary school classrooms. *The Journal of the Acoustical Society of America, 123*(4), 2064.

Shendell, D. G., Prill, R., Fisk, W. J., Apte, M. G., Blake, O., & Faulkner, D. (2004). Associations between classrooms' CO_2 concentrations and student attendance in Washington and Idaho. *Indoor Air, 14,* 333–431.

Taylor, A. (1993). The learning environment as a three-dimensional textbook. *Children's Environments, 10*(2), 170–179.

Taylor, A. (2009). *Linking architecture and education: Sustainable design of learning environments.* Albuquerque: University of New Mexico Press.

US Green Building Council's Center for Green Schools. (n.d.). Retrieved from www.centerforgreenschools.org/green-schools

Williams, I. (2020, September). Creating adaptive capacity to design a more sustainable tomorrow. *Green Schools Catalyst Quarterly,* 38–45. Retrieved from https://catalyst.greenschoolsnationalnetwork.org/gscatalyst/september_2020/MobilePagedReplica.action?pm=2&folio=1#pg1

1. LEAD 2. DESIGN 3. MAINTAIN & OPERATE 4. TEACH 5. LEARN 6. PLAY 7. MODEL 8. PARTNER 9. START SMALL

STRATEGY

3

Maintain and Operate

Operate and Maintain Healthy, Safe, Sustainable Learning Environments

Living systems-minded school leaders, together with their facilities colleagues, manage healthy, safe, and sustainable learning environments in ways that reduce energy, conserve natural resources, and minimize waste. Sustainability-focused facility maintenance and operations routines support occupants' daily learning, their general health, and their overall well-being, at the same time they minimize or eliminate negative environmental impacts. These sustainable maintenance and operation routines also provide opportunities for utilizing the school facility as a three-dimensional textbook (Kong, Rao, Abdul-Rahman, & Wang, 2014; Nair & Fielding, 2005; Taylor, 1993).

This strategy outlines various steps living systems-minded school leaders take as they (1) work with facility colleagues to manage healthy, safe, and sustainable schools; (2) seek energy efficient alternatives; (3) conserve resources; (4) utilize green operations and maintenance systems and routines to engage students with the three-dimensional textbook they inhabit each day; and (5) marshal the public will to provide high-quality, sustainable learning environments for all children.

Work With Facilities Professionals to Maintain Healthy, Safe, Sustainable Schools

Living systems-minded school leaders build bridges between the historically divided curriculum and facilities sides of the academic house, never underestimating the capacity, or the desire, of maintenance and operations colleagues to lend their expertise. Curriculum and facilities professionals employ their collective imagination, transforming a laundry list of mundane maintenance and operations tasks into an integrated plan for utilizing the school building as a three-dimensional textbook (Kensler & Uline, 2017). Student involvement in maintaining indoor air quality, controlling moisture, and keeping things clean, "give students opportunities to try new, sustainable behaviors" (Higgs & McMillan, 2006, p. 45).

Ensuring High-Quality, Indoor Air

In the McGraw-Hill Construction (2013) report, 88% of K–12 respondents considered enhanced occupant health and well-being to be a primary catalyst for their greening efforts, roughly equivalent to energy use reductions and operating cost savings. Further, products and practices that improve indoor environmental quality (IEQ) were deemed essential to achieving this goal, with 87% of respondents ranking indoor air quality (IAQ) practices as highly important.

Integrated design decisions, which combine operable windows with well-designed mechanical systems, ensure an abundance of fresh air within classrooms (National Research Council, 2007). Designers, well versed in sustainable practices, urge school leaders to "think simplicity, not high tech, . . . starting with windows that open and close" (LPA Architects, 2009, p. 23). In fact, research suggests teachers highly value influence and control over their physical settings, especially the ability to open windows and allow the circulation of fresh air (Uline, Tschannen-Moran, & Wolsey, 2009). Careful placement of fresh air intakes also limits the intrusion of exhaust and other pollutants generated by motor vehicles and equipment that frequent school sites. Regular maintenance of ventilation pathways control pollutants and moisture from incoming air.

Green school guidelines call for the elimination of gas-fired pilot lights and discourage the use of fossil fuel-burning equipment indoors. Guidelines also call for dedicated exhaust systems for spaces that might house chemicals, including cleaning equipment and supply storage areas, photography labs, copy/print rooms, and vocational spaces (National Research Council, 2007).

Even before school construction is completed and the building is occupied, dust and outgassing contaminants, introduced by conventional construction materials, influence IAQ in negative ways. Sustainably sourced, naturally made building materials, including flooring, acoustic ceiling tiles, insulation, signage, and wall paneling eliminate these contaminants. As construction progresses, daily HEPA vacuuming of all soft surfaces, replacement of all filters at the completion of construction, and 28 days of continuous flushing of the building with outside air prior to occupancy are all proactive steps to maintaining clean indoor air (National Research Council, 2007).

Employing the Three-Dimensional Textbook

The EPA Tools for Schools Action Kit includes a Teacher's Classroom Checklist that provides opportunity for engaging students in promotion of high-quality indoor air. Students might assist in completing the checklist, brainstorming strategies, and assuming appropriate levels of responsibility for maintaining general classroom cleanliness, managing animals in the classroom, reducing moisture sources, and taking other preventative actions to ensure healthy indoor air. Students might also examine construction materials within the classroom and sort by type: renewable, nonrenewable, recyclable, natural, man-made to learn more about their potential benefits and/or harm as it relates to IAQ (Taylor, 2009, p. 196). These learning activities build students' observational, research, and problem-solving skills. They also provide students opportunities to assume responsibility as active members of their school's maintenance team, contributing to the health of their school's ecosystem.

Controlling Moisture

The building envelope physically separates the interior and exterior of a building and includes the foundation, walls, floors, roofs, fenestrations, and doors. Any building assembly exists in dynamic relation to its external environment (National Research Council, 2007). Therefore, integrated design decisions related to building siting, design, and materials increase our ability to control moisture. Building scientists conduct "source-path-driving force" analyses to determine the source of the moisture, the path it follows, and the force that drives the moisture along the pathway. Control of at least one of the three elements allows for effective control of moisture, with control of more than one element providing valuable redundancy (National Research Council, 2007).

Certain construction materials, such as masonry walls, have the capacity to store moisture and then dry without harmful effects (National Research Council, 2007). Well-designed drainage systems and HVAC condensation drainage systems also prevent water accumulation. On the exterior of the building, eaves and roof overhangs direct rainwater away from building walls; at the same time, ground, sloped away from the building, carries runoff out from the walls and foundation (Freed, 2010).

Building operations and maintenance strategies play a key role in avoiding moisture and diagnosing moisture sources when they occur. Keeping site irrigation to a minimum, using walk-off grills and mats to prevent rain and snow from entering the building, inspecting regularly for building leaks, and storing construction materials in dry and well-ventilated areas all constitute sound moisture control strategies.

Keeping Things Clean in Safe and Healthy Ways

Traditional cleaning supplies adversely affect the environment, the occupants of schools, and, in particular, the people employed to clean them (Gelfand,

2010). To avoid these hazards, living systems-minded school leadership teams increasingly choose environmentally responsible cleaning supplies and procedures. Since 2005, 11 states, including Connecticut, Hawaii, Illinois, Iowa, Maine, Maryland, Missouri, Nevada, New York, Vermont, Mississippi, and the District of Columbia, have enacted laws addressing green cleaning in schools. Recent findings suggest these laws have potential to raise awareness and encourage use of green cleaning products and practices in schools (Arnold & Beardsley, 2015).

Mechanical and structural systems are influenced by occupant use over time and by operations, maintenance, repair, and cleaning practices (National Research Council, 2007). We are wise to keep things simple and understandable (Edwards, 2006; Gelfand, 2010; LPA Architects, 2009). In this way, facilities professionals can better learn systems and procedures; trust their effectiveness; and teach principals, teachers, and students. As an example, EUSD instituted the lotus® PRO High Capacity (https://info.waxie.com/products/chemical-free-cleaning/lotus-pro-chemical-free-cleaning-system) cleaning system which transforms ordinary tap water into an effective all-natural commercial cleaner by infusing the water with ozone (See Figure 0.1. Encinitas Union School District Green Initiatives, beginning on p. 22). The resulting cleaning agent replaces traditional chemical cleaners, deodorizers, and sanitizers. Dr. Baird described the learning curve experienced by district custodians.

> We instituted a water cleaning system that changes the chemical alignment of tap water so that it actually works without cleaning agents. We had to teach our custodians. We had to get their buy-in, because it didn't smell like bleach. Part of it was convincing them, "Yes, this really does clean." We actually did science experiments. The custodians wiped down one desk with bleach and another desk with the water. Then they checked the level of contaminants left behind on both desks. They finally got it! Now the custodians are our biggest fans of the system, because it's much easier and safer.

Operate Energy Efficient Schools

Along with energy efficient lighting, heating and cooling, and ventilation systems, careful siting and building orientation allow living systems-minded school leaders to "[take] advantage of the site's free gifts" (LPA Architects, 2009, p. 24). Mindful building orientation allows for maximum use of natural daylighting and increased opportunities for natural ventilation via prevailing breezes (LPA Architects, 2009).

Utilizing Renewable Energy Technologies

Living systems-minded school leaders are also utilizing renewable energy technologies to power their schools.

- Geothermal systems draw heat from the earth and expel heat back into the earth to heat and cool the building.

- Biomass systems utilize biological waste materials to generate electricity, while photovoltaic systems utilize solar panels on roofs, shade surfaces, and facades.

(LPA Architects, 2009)

- K–12 schools are among the fastest adopters of solar power in the United States.

A 2017 report by the Solar Foundation and the Solar Energy Industries Association revealed that, initially, school districts adopted solar for the educational or symbolic value of small systems, but, today, schools are tapping solar on a much larger scale. According to the authors of the report, 5,489 solar-equipped K–12 schools now yield 1.4 million megawatts, while offsetting an estimated 1,030,873 million metric tons of carbon dioxide emissions annually, equivalent to the greenhouse gas emissions from nearly 221,000 cars or the carbon sequestered by 27 million trees" (The Solar Foundation, 2017. Available at: www.thesolarfoundation.org/solar-schools). In fact, nearly 3.9 million, or 7.3% of U.S. students, now attend a school powered with solar energy.

As mentioned earlier, district bond dollars were used to install solar panels on all nine campuses within Encinitas School District (see Figure 0.1. Encinitas Union School District Green Initiatives, beginning on p. 22). Dr. Baird reported substantial annual savings, underscoring the immediate effect of these savings on the school district's operating budget, as monies could be redirected from utilities costs to teaching and learning functions.

We were going to spend bond dollars to put solar on all of our campuses. We couched it in two ways. One, "We're going to do the right thing for the environment", and, two, "We're going to save money for the district in the long run." There were some people who didn't think it was the right way to go. They were thinking of home solar, criticizing the nature of a 10-year return on investment. In response, we described the difference for schools, where you're spending bond dollars that can only be expended on capital improvements, but you're getting the money back in operational dollars that would have been used for utility costs. So, day one you start seeing that return on investment. We have estimated it's somewhere in the neighborhood of $800,000, to potentially $1 million, annual savings.

Solar installations offer opportunities to enhance STEM (science, technology, engineering, and math) learning with hands-on, real-world tools. This addition to the three-dimensional textbook provides students access to technology that helps prepare them for a future in the solar industry, which boasts the fastest-growing occupation in the nation (U.S. Department of Labor, Bureau of Labor Statistics, Occupational Outlook Handbook, last modified September 4, 2019, www.bls.gov/ooh/fastest-growing.htm as referenced in Generation180, 2019, p. 5). Solar panels, installed at all nine EUSD campuses, provide such learning opportunities (See Figure 3.1).

Employing the Three-Dimensional Textbook

In green schools, students can learn about energy stewardship by collecting and comparing electricity use data from schools across their district. They can trace the path of their school power, from "source to outlet", experiencing alternative energy and energy con-servation in a hands-on manner (Taylor, 2009, p. 202). Students cal-culate each school's daily energy consumption by multiplying the watts times the hours used to determine watt-hours. They might also divide by 1,000 to verify the kilowatt-hours, in the same manner utility companies compute charge for usage. Comparisons across schools provide students opportunities to practice data collection, analysis, and reporting skills, focusing attention on real world concerns, i.e., their school districts' bottom line.

Figure 3.1 Solar panels are installed at all nine EUSD campuses.

Insulating Well

Insulation is "a simple and vital means of conserving energy", as it "keeps hot air out and cool air in during summer and warm air in and cold air out during winter" (LPA Architects, 2009, p. 37). Insulation begins with the building frame. "A 2x4-framed wall . . . will have an insulating value of R13 using conventional fiberglass batt insulation. A 2x6-framed wall, however, accommodates R19 insulation, [providing] 50 percent more insulation" (p. 37). As you increase the R-value of insulation you increase its effectiveness. It should be noted that soy-bean-based or cotton insulation avoids toxins (LPA Architect, 2009).

Thermally broken windows, with separator material between inner and outer window frame, prevent temperature transfer. High thermal-mass materials, such as concrete and stone, can also help maintain a consistent internal temperature. Outside the building, earth berms and green screens, placed along exterior walls, provide further insulation from heat gain, reducing HVAC use and energy consumption. These landscape design elements also provide green transitions.

Maximizing Daylight

Daylighting can cut a school building's lifetime energy costs by 30–70%, with diffuse light provided by means of baffles, roof monitors, skylights, and clerestories (Olson & Kellum, 2003). Operable windows with low-E laminated glazing minimize the ultraviolet and infrared light that can pass through glass without compromising the amount of visible light that is transmitted, stopping the direct transfer of heat but not sunlight. Abundant natural light reduces need for artificial light sources, which are responsible for up to 60% of a school building's energy consumption (Gelfand, 2010). A coordinated system that utilizes daylight in combination with indirect and energy efficient artificial lighting options, such as advanced florescent lamps and ballasts, light emitting diodes (LEDs), and/or high intensity discharge (HID) go a long way to providing high-quality lighting and reducing overall costs. Classroom dimming systems automatically dim or shut down indirect and direct artificial lighting when sufficient daylight is present. Occupancy sensors shut off lighting when no one is present.

As an additional system for harvesting daylight on their campuses, EUSD opted to install solar tubes where possible (See Figure 0.1. Encinitas Union School District Green Initiatives, beginning on p. 22). Figure 3.2 shows solar tubes installed on the roof of one EUSD school building. These tubes contain prismatic lenses that transmit and diffuse any available light throughout the day, regardless of the sun's angle. In addition to the lens at the top, the tube itself is highly reflective. As the light enters through the prismatic lens, it bounces about the tube and is distributed into the classroom through an additional lens at the bottom. The tubes are also louvered,

Figure 3.2 Solar tubes are installed where possible on all nine EUSD campuses.

allowing teachers to shut the sun off. This feature provided EUSD students yet another opportunity to join their school's maintenance and operation teams. Dr. Baird explained,

> We had to remember, you've got to bring people along slowly. The solar tubes have a manual switch. Teachers may close them so they can show something on their screen, but then they forget and leave them closed. We've been fighting this battle for a while. Students are your best bet here, always. The students have gone around and checked to see who's got their tubes open and who has them closed. They've been our biggest advocates, leading information campaigns. In some of our schools, each classroom assigns an energy czar to ensure the solar tubes are open when they should be open.

Employing the Three-Dimensional Textbook

Informed students become green school leaders' best monitors of daytime use, reminding teachers and peers to turn off unnecessary lights when not in use and turn down lighting levels when possible. They can go a step further to chart the savings that results from responsible artificial and natural light usage. Students might also describe all the artificial light sources throughout their school, exploring the differences between high-pressure sodium, fluorescent, full spectrum (Taylor, p. 191). These activities might provide the necessary evidence for writing persuasive essays, making the argument for natural light as a cost-effective and health-conscious alternative to artificial light sources.

Heating and Cooling Efficiently

Low-energy heating and cooling methods reduce energy costs and maintain comfortable, consistent temperatures for building occupants. High-performance heating, ventilation, and cooling systems share common characteristics such as:

- Over-size condenser coils that lead to higher equipment efficiencies,
- Water-cooled, as opposed to air-cooled, condensers,

- Mechanical systems that recover unwanted heat or water in one part of the building and transfer it to another part that requires heat,
- Extended periods of economizer cycle operation.

(Gelfand, 2010, p. 145)

Efficiency is increased and indoor pollutants minimized through regular maintenance of heat, ventilation, and air conditioning (HVAC) systems, utilizing high-efficiency filters and ducted returns.

Employing the Three-Dimensional Textbook

Students might conduct a walking tour of their school, measuring building temperature in different locations. They could diagram the human circulatory system and compare it to their schools building systems (Taylor, 2009, p. 199). This learning experience provides students the opportunity to compare a familiar natural system with a mechanical one, sparking larger conversations about all the ways knowledge of natural systems has potential to inform improvements in man-made systems.

Monitoring Building Performance

In order to realize and maintain a green school's potential health and prod- uctivity benefits, living systems-minded leaders are wise to utilize monitoring and diagnostic feedback regarding building performance (National Research Council, 2007). The ability to effectively track the long-term savings realized through sustainable practices supports the case for future investments in green building and retrofits (McGraw-Hill Construction, 2013).

Employing the Three-Dimensional Textbook

Monitoring systems, combined with interactive energy dashboards, also provide opportunities for learning as students observe, analyze, and report results. EPA's ENERGY STAR Portfolio Manager is a web-based utility tracking tool that allows eligible facilities (including K–12 schools) to record, track, and benchmark their energy and/or water performance against that of similar facilities across the country. EPA's ENERGY STAR energy performance scale indicates how efficiently buildings use energy on a 1-to-100 scale. An ENERGY STAR energy performance score of 50 indicates average energy performance while an ENERGY STAR Score of 75 or better indicates top performance. Energy dashboards provide students the opportunity to monitor building performance and report results, learning important math and science skills, at the same time functioning as the energy conscience of their school community. (See Discovery School's Energy Dashboard, available publicly at https://discovery. apsva.us/tour-energy-dashboard/)

Conserve Resources

Green schools are designed to consume fewer natural resources and create a harmonious relationship between the building and its natural surroundings (LPA Architects, 2009). These schools are learner-centered; at the same time, they respect and protect the natural world.

Reducing Water Use

School campuses are often large enough to influence water quality and quantity beyond their boundaries (Herrington, 2010). Along with water-conserving plumbing fixtures in kitchen facilities, restrooms, and gymnasium locker rooms (including low-flow toilets and aerators on the sinks), green roofs and rain gardens also reduce water runoff, becoming a piece of local habitat (Freed, 2010). Grass, gravel, and resin paving; unit pavers; and porous asphalt all allow water to percolate and/or plants to grow, while also reducing heat islands (Herrington, 2010, p. 171).

Green school leaders are incorporating bioswales into school grounds, planted with native species that help to purify water runoff. Drought tolerant grasses require less irrigation, and water recapture systems for irrigation reduce water use, moving in the direction of net zero (Herrington, 2010). Strategies such as avoiding midday watering, zoning irrigation by plant location and type, maintaining heads and filters, utilizing smart controllers, and using low-volume irrigation for gardens, trees, and shrubs, together provide an integrated strategy for conserving water (Eley, 2006; Gelfand, 2010).

Employing the Three-Dimensional Textbook

All fifth-grade students in the EUSD completed a unit on water scarcity. First students learn how little fresh water is currently available worldwide, and then they learn about filtration, both biological and physical. They conduct a classroom experiment, observing how plant roots act as biological filters to purify water, writing their procedures and findings in individual journals, seen in Figure 3.3.

Students also learn about physical filters and create their own four-layer water filters (See Figure 3.4). They rate the dirty water, and then rate the effectiveness of their filters in cleaning it, making additional attempts to improve upon their designs, based on their results (See Figure 3.5).

As part of this unit, students engage in several local field trips, first visiting a local farm to observe regenerative agriculture practices that conserve water. They also visit a local lagoon and natural watershed to understand how the watershed cleans the water as it goes into the ocean. They visit Encinitas's reclamation plant to see how their community reclaims and reuses water for irrigation and other purposes. Finally, the students learn about different water pumping systems. Teams of students design pumps that will pump and filter water from the Farm Lab bioswale, which fills with thousands of gallons of water. This water will be used for irrigation on the farm.

Figure 3.3 Students capture observations in their Experiment Journals.

Figure 3.4 Student teams build four-layer physical water filters.

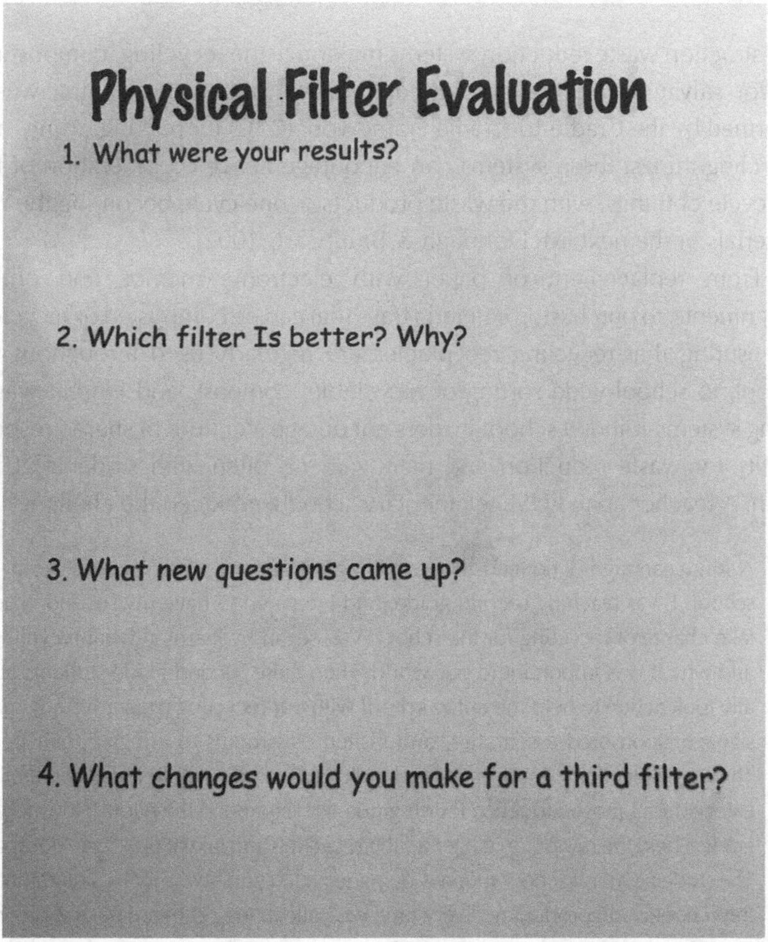

Physical Filter Evaluation

1. What were your results?

2. Which filter Is better? Why?

3. What new questions came up?

4. What changes would you make for a third filter?

Figure 3.5 Each student reports the outcomes of their group experiment.

Preserving and Protecting Habitats

The reuse of existing school sites, in combination with compact school designs, help to preserve undeveloped, open spaces. In addition, living systems-minded school leaders are taking account of their school sites as natural systems, respecting and preserving wetlands and other existing habitats as laboratories for learning.

Manage Waste

Construction waste reduction systems maximize the recycling, composting, and/or salvaging of construction, demolition, and land-cleaning waste. Informed by the Cradle-to-Cradle Framework (C2C) for making, using, and recycling things, these systems can encourage active consideration of the life cycle of things, with the waste products of one cycle becoming the raw materials of the next (McDonough & Braungart, 2002).

From replacement of paper with electronic memos and online assignments, to purchasing cafeteria trays that can be composted or recycled, to ensuring that recycling receptacles are regularly used throughout the school, to school-wide sorting of recyclable, compost, and landfill waste, living systems-minded school leaders encourage a culture of shared responsibility for waste reduction and management, often with students at the helm. A teacher at an EUSD elementary school embraced the challenge.

> A few years ago, I noticed that recycling just wasn't what it could be at our school. I was teaching second grade, and I decided to have my second-graders take charge of recycling for the school. We began by learning what recycling is and why it was important to our world. Then these second-grade students actually took action to help the entire school with our recycling program. They made signage, sponsored assemblies, and visited classrooms to educate their peers. Through their efforts, each grade level took responsibility for their daily collection. Later, when I moved to teach fourth grade, we expanded the effort. Now, fourth-graders help the custodian gather all the recycling and take it out to the dumpster. The students monitor how much is being recycled and how much contamination the school is still producing. Every time we collect, they gather data and record it on a Google form so they can strategize ways to gauge progress. Each grade level hallway has a different group responsible for their classrooms' collection efforts. Next week, representatives from these groups will come together to notice and analyze trends and design strategies for improvement.

The teacher also shared how fourth-graders expanded their sphere of influence, demonstrating how meaningful academic work builds students' capacity to be active agents of change within their communities.

> We did a project last year involving the Encinitas YMCA. They were interested in reducing their carbon footprint, and so they invited my class to document what they were recycling and what they weren't. Upon completion of the audit, the students presented their findings and gave them different solutions

on how to become better recyclers. Already you can go to the Y and see the one-to-one stations they've set up. They are saving a lot of money, and I think they're feeling good that they're actually doing a lot more recycling. The whole class presented their work to our school board. We invited the YMCA leaders and the parents. They all had a piece of that presentation. We also had three students go to the YMCA and present to their board. Here they were presenting in front of audiences of adults, and adults that can make a difference!

Employing the Three-Dimensional Textbook

As mentioned in Chapter 1, EUSD developed the SCRAP Cart (Separate, Compost, Reduce, And Protect) in order to reduce land-fill waste, facilitate composting, and encourage lunchtime recycling district-wide (See Figure 0.1. Encinitas Union School District Green Initiatives, beginning on p. 22). The SCRAP Cart is used to teach students how to properly sort their lunchtime waste for composting, recycling, and landfill. Since the introduction of the carts in 2012, lunchtime waste has been reduced at each school by over 80%, saving the school district over $40,000 every year (See Figure 3.6).

Figure 3.6 A waste diversion program is implemented district-wide via food SCRAP composting.

Marshal the Public Will

In their recent State of Our Schools report, the 21st Century School Fund, in partnership with the U.S. Green Building Council, and the National Council on School Facilities, estimated it will take approximately $145 billion per year to maintain, operate, and renew school facilities in the United States so they provide healthy and safe 21st-century learning environments for all children (Filardo, 2016). The nation's current system for funding educational infrastructure leaves too many school districts unable to provide adequate and equitable places for learning.

> Comparing historic spending against building industry and best-practice standards for responsible facilities stewardship, we estimate that national spending falls short by about $8 billion for M&O and $38 billion for capital construction. In total, the nation is under spending on school facilities by $46 billion.
>
> (Filardo, 2016, p. 4)

As a nation, we have allowed too many of our schools to fall into a state of disrepair we would not accept in the places where we work, eat, recreate, or shop. Too often, chronic funding shortages place building maintenance and repairs low on the list of school district priorities, resulting in the degradation of basic heating, ventilation, lighting systems, (Kats, 2006), particularly in school districts serving high concentrations of low-income students (National Center for Education Statistics, 1995). These districts disproportionately draw more from general operating funds for operations and maintenance than districts serving higher income students, leaving fewer dollars for education programs (Vincent & Jain, 2015).

As we seek to address the needs of students and communities in the 21st century, we have opportunity to reverse this trend and marshal the public will to meet industry standards and provide high-quality, sustainable schools for *all* our children. A recent study of new schools, built in the United States between 2000 and 2014, reported some progress in this regard (Zhao, Zhou, & Noonan, 2019). Researchers found that 2,159 (or 6.4%) of newly constructed schools were green, that is, designated by the US Green Building Council as LEED-certified or LEED-registered (in the process of being certified). This designation indicates adherence to strict environmental and energy standards, using the U.S. Green Building

Council's LEED (Leadership in Energy and Environmental Design) rating system. Researchers further examined the demographic characteristics of the students who attend these schools, as well as the demographics of communities within which the schools are situated, discovering that the new green schools serve lower-income and minority family and children at higher rates than their more affluent counterparts, thus advancing the goals of environmental justice for these under-represented student populations (Zhao et al., 2019). The authors suggest that public policy could "encourage, reinforce, or formalize [this] nationwide pattern that has formed seemingly without national regulation or much deliberate, systemic policy" (Zhao et al., 2019, p. 2236).

Conclusion

Operating and managing schools with a living-systems frame of mind increases the likelihood that our schools will support the whole child, not only creating the conditions for their academic success, but also addressing their needs in terms of physical, emotional, social, and cognitive well-being. As we sharpen and extend our focus to include the natural dimensions of school life (human and otherwise), the facility-related job functions of principals, custodians, plant managers, district shop foremen take on new importance and present exciting possibilities for reducing a school's carbon footprint, for saving precious dollars, for enriching students' learning experiences, and for meeting our children's needs, now and into the future.

 Leadership Design Challenges

1. **Design your school's (school district's) version of a SCRAP Cart.** Invite students to play an active role in the design and implementation of this lunchtime recycling tool. Embrace this project as a teaching tool, a cost-saving strategy, and a means to kickstart a living-systems, sustainability-focused frame of mind across your school/school district. Be sure to confirm that your local municipality supports recycling efforts. If not, consider engaging students in a community-wide effort to encourage recycling.

2. **Develop a schoolyard wildlife habitat.** The National Wildlife Federation (NWF) assists schools in developing outdoor classrooms called Schoolyard Habitats®, where educators and students learn how to attract and support local wildlife. The program sponsors a listserve for organizations working on schoolyard improvement. Web pages for National Wildlife Foundation (www.nwf.org/), the Evergreen Foundation (www.evergreen-foundation.com/), and Project Wild (www.projectwild.org/) provide information and links to other similar projects (Rivkin, 1997).

   a. The Schoolyard Habitat Program has certified schoolyards in every U.S. state and two territories, with international sites in Thailand, Italy, and the United Kingdom. These wildlife habitats become places where students learn about wildlife species and ecosystems, while at the same time honing academic skills and nurturing creativity (National Wildlife Federation www.nwf.org/How-to-Help/Garden-for-Wildlife/Schoolyard-Habitats.aspx).

   b. Start small with projects such as butterfly gardens, bird feeders and baths, tree planting, sundials, weather stations, and native plant gardens. Larger projects, often initiated in concert with new construction, include wetlands, nature trails, meadows, stream restoration, shelters for small animals, and large vegetable gardens (Rivkin, 1997). In addition to developing and caring for schoolyard habitats, students' sense of ownership and stewardship might be enriched through a process of identifying, photographing, and archiving the native plants and animals thriving within their schoolyard habitat.

 Learning From Living Systems-Minded Trailblazers

Middlesex County Public Schools (MCPS) is a small, conservative-leaning, rural school district, located on the Virginia coast. Forty-one percent of MCPS students qualify for free and reduced-price lunches. MCPS was the first school district in the region to install solar panels on site and the first school district in Virginia to meet 100% of its schools' electricity needs with solar energy. A November 2019 report, entitled "Powering A Brighter Future", released by Generation180 (a Virginia-based nonprofit working to inspire and equip people to take action on clean energy), tells the story of Greg Harrow's, MCPS's Director of Operations & Transportation, journey from solar skeptic to solar advocate (Generation180, 2019). Initially, Harrow was reluctant to undertake a project that might increase workload for his small facilities team. He was also wary of the related costs.

Power purchase agreements (PPAs) are the primary method Virginia schools use to finance solar installations. Since 2014, PPAs have accounted for nearly 90% of K–12 solar school installations nationwide (Generation180, 2019, p. 9). Under a PPA, a third-party system purchases, owns, and maintains the solar panels. School districts agree to buy the electricity produced by the system for the length of the agreement, often 25 or more years. PPAs make it easier for school districts to install solar with little-to-no upfront investment or ongoing maintenance costs. In addition, the school district typically pays a lower electricity rate than it previously paid the utility, resulting in immediate energy cost savings. For MCPS, this means a cumulative savings of $4.74 million over 25 years. Harrow now boasts the district's savings to frequent visitors who come to learn about the district's energy transformation.

References

Arnold, E., & Beardsley, E. R. (2015). *Perspectives on implementation and effectiveness of school green cleaning laws*. Washington, DC: U.S. Green Building Council.

Edwards, B. W. (2006). Environmental design and educational performance. *Research in Education, 76*, 14–32.

Eley, C. (2006). High performance school buildings. In H. Frumkin, R. J. Geller, & I. L. Rubin (Eds.), *Safe and healthy school environments* (pp. 341–350). New York: Oxford University Press.

Filardo, M. (2016). *State of our schools: America's K-12 facilities 2016*. Washington, DC: 21st Century School Fund, U.S. Green Building Council, Inc., & the National Council on School Facilities.

Freed, E. C. (2010). Building structure and envelope. In L. Gelfand (Eds.), *Sustainable school architecture: Design for primary and secondary schools* (pp. 111–136). Hoboken, NJ: John Wiley & Sons, Inc.

Gelfand, L. (2010). *Sustainable school architecture: Design for primary and secondary schools*. Hoboken, NJ: John Wiley & Sons, Inc.

Generation180. (2019). *Powering a brighter future: A report on solar schools in Virginia*. Charlottesville, VA: Generation180.

Herrington, S. (2010). Landscape and site design. Contributor to L. Gelfand, *Sustainable school architecture: Design for primary and secondary schools* (pp. 163–196). Hoboken, NJ: John Wiley & Sons, Inc.

Higgs, A. L., & McMillan, V. M. (2006). Teaching through modeling: Four schools' experiences in sustainability education. *The Journal of Environmental Education, 38*, 39–53. doi:10.3200/JOEE.38.1.39-53

Kats, G. (2006). *Greening America's schools: Costs and benefits*. Retrieved from www.usgbc.org/showfile.aspx?DocumentID-2908

Kensler, L. A. W., & Uline, C. L. (2017). *Leadership for green schools: Sustainability for our children, our communities, and our planet*. New York: Routledge/Taylor & Francis Group.

Kong, S., Rao, S., Abdul-Rahman, H., & Wang, C. (2014). School as 3-D text-book for environmental education: Design model transforming physical environment to knowledge transmission instrument. *Asia-Pacific Education Researcher, 23*(1), 1–15. doi:10.1007/s40299-013-0064-2

LPA Architects. (2009). *Green school primer*. Victoria, Australia: The Images Publishing Group.

McDonough, W., & Braungart, M. (2002). *Cradle to cradle: Remaking the way we make things*. New York: North Point Press.

McGraw-Hill Construction. (2013). *New and retrofit green schools: The cost benefits and influence of a green school on its occupants*. Bedford, MA: McGraw-Hill Construction.

Nair, P., & Fielding, R. (2005). *The language of school design: Design patterns for 21st century schools*. Minneapolis, MN: Designshare, Inc.

National Center for Education Statistics (NCES). (1995). *Disparities in public school district spending 1989–90*. Washington, DC: US Department of Education.

National Research Council. (2007). *Green schools: Attributes for health and learning*. Washington, DC: The National Academies Press.

Nolet, V. (2016). *Educating for sustainability*. New York: Routledge.

Olson, S. L., & Kellum, S. (2003). *The impact of sustainable buildings on educational achievement in K–12 schools*. Madison, WI: Leonardo Academy, Inc.

Rivkin, M. (1997). The schoolyard habitat movement: What it is and why children need it. *Childhood Education Journal*, *25*(1), 61–66.

The Solar Foundation. (2017). *Brighter: A study on solar in U.S. schools*. Retrieved from www.thesolarfoundation.org/solar-schools

Taylor, A. (1993). The learning environment as a three-dimensional text-book. *Children's Environments*, *10*(2), 170–179.

Taylor, A. (2009). *Linking architecture and education: Sustainable design of learning environments*. Albuquerque: University of New Mexico Press.

Uline, C., Tschannen-Moran, M., & Wolsey, T. D. (2009). The walls still speak: The stories occupants tell. *Journal of Educational Administration*, *47*(3), 400–426. doi:10.1108/09578230910955818

Vincent, J. M., & Jain, L. S. (2015). *Going it alone: Can California's K–12 school district adequately and equitably fund school facilities?* Berkley, CA: Center for Cities and Schools. Retrieved from http://citiesandschools.berkeley.edu/uploads/Vincent__Jain_2015_Going_it_Alone_final.pdf

Zhao, S., Zhou, S., & Noonan, D. S. (2019). Environmental justice and green schools-assessing students and communities' access to green schools. *Social Science Quarterly*, *100*, 2223–2239. doi:10.1111/ssqu.12715

ACTION

2

Take Students Outside

Strategies for reconnecting students with nature might begin simply with opening the doors to the outside and reintroducing recess. Emerging research suggests improved student behavior, learning focus, and academic performance follow daily recess, unstructured play in the outdoors. Beyond recess, academic learning can also occur productively while deeply embedded in the outdoors. Students learn basic content, in addition to gaining deep insight into how the world works as an integrated, interdependent whole. Numerous recent reviews of research demonstrate that contact with nature is associated with overall health and well-being, including specific aspects of emotional, physical, social, and cognitive well-being. The evidence is substantial—children benefit from contact with nature.

STRATEGY

4

Teach
Prepare Teachers to Teach in Nature

Living systems-minded school leaders provide ongoing, job-embedded opportunities for teachers to learn the requisite knowledge and skills for facilitating student learning in nature. They challenge teachers to move outside their classroom comfort zone and provide the necessary resources and supports to ensure teachers' success in doing so. As leaders confidently model sustainability-related behaviors, dispositions, and habits of mind, they help teachers conquer their fears about enacting innovative teaching practices in and with nature. Living systems-minded school leaders explicitly demonstrate how teaching in nature supports teachers' efforts to deliver rigorous, integrated academic content. They expect nothing less.

This strategy outlines various steps living systems-minded school leaders take as they (1) provide time and resources to develop and implement interdisciplinary, learner-centered, problem- and place-based learning experiences that happen in nature; (2) invent systems and structure to support the work of preparing teachers to teach in nature; (3) model effective nature-based teaching practices; (4) introduce teachers to their green school facility as a three-dimensional textbook.

Provide Time and Resources to Develop Interdisciplinary, Learner-Centered, Problem- and Place-Based Learning Experiences That Happen in Nature

Nature-based, sustainability-focused teaching and whole school approaches to delivering such instruction rests in a set of values, knowledge, skills, and ways of thinking that have potential to energize teachers and build their sense of efficacy for addressing the diverse needs of learners. It is incumbent upon school leaders to provide teachers the necessary resources to develop and implement the interdisciplinary, learner-centered, problem- and place-based learning experiences upon which nature-based learning depends. As teachers engage in this work, they discover how these authentic, real-world experiences motivate their students to learn complex concepts and skills. In fact, it may be students who provide the most persuasive arguments for investing time and effort in learning these methods as they "look to their teachers for answers and assurances about a safe, healthy, and sustainable future" (Bauermeister & Diefenbacher, 2015, p. 326). Our children may likely be the ones to "show [us] how sustainability could and should be seen as an urgent social imperative and critical context for twenty-first century learning" (McClam & Diefenbacher, 2015, p. 129).

Introducing Underlying Ideas and Values

The Earth Charter, formally instituted at the Hague Peace Palace in 2000 and signed by 6,000 governmental agencies, NGOs, businesses, universities, schools, and religious and youth groups, is now considered "a global consensus statement on the core principles of sustainability" (Nolet, 2016, p. 51). The Charter speaks directly to educating children and youth in ways that "empower them to contribute actively to sustainable development" (The Earth Charter International, 2015). Begun in 2005, The United Nations' Decade of Education for Sustainable Development (UNDESD) sought large-scale changes in our educational systems aimed at promoting this sort of education. Teacher preparation and professional development were among the primary foci for UNDESD work (Wals, 2009). As living systems-minded school leaders begin to examine how best to institute sustainability-focused curriculum and instruction within their schools and school districts, they might spend some time exploring related documents and websites as a foundation for their ongoing inquiry. Table 4.1 summarizes commonly utilized terms and sources.

Table 4.1 Terms Associated with Nature-based Teaching and Learning for Sustainability

Environmental Education (EE)	Engages students in study of the environment to 'encourage behavior change and action' (Thomas, 2005). EE employs hands-on activities and relevant subject matter to engage students and encourage participation. EE is "a creative and dynamic process in which pupils and teachers are engaged together in a search for solutions to environmental problems." (http://unesdoc.unesco.org/images/0010/001056/105607e.pd) (Riordan & Klein, 2010)
Education for Sustainable Development (EfSD)	Prepares learners to make informed decisions and take responsible actions aimed at environmental integrity, economic viability, and a just society. EfSD is interdisciplinary and holistic, embedded in the whole curriculum, and values based. It emphasizes open-ended, generative thinking, critical thinking, problem solving, participatory decision making and systems thinking. EfSD is based on local context but connected to global issues, is culturally responsive and learner centered. (https://en.unesco.org/themes/education-sustainable-development)
EcoJustice Education	Teaches protection of living systems and community well-being through examination and response to what degrades them. EcoJustice Education recognizes the importance of biological and cultural diversity, and the

(Continued)

Table 4.1 (Continued)

	need to make decisions on behalf of all who will be most affected, including future generations and the more than human world. Recognizes and revalues diverse commons-based practices, traditions, and knowledge from cultures and communities worldwide. (Lowenstein, Martusewicz, & Voelker, 2010)
Education for Sustainability (EfS)	Seeks to equip learners to deal with the challenges that arise from the interconnectedness of environment, culture, society, and economy that typifies life in the 21st century. EfS helps learners develop new ways of thinking, collaborating, and solving problems. Learning focuses on local concerns, interpreted through larger global perspectives. EfS is interdisciplinary, holistic, and embedded across the curriculum. It is culturally responsive and values-based. Employs a variety of methods and pedagogies, with particular attention to learner-centered strategies. (Nolet, 2016)
Whole School Sustainability	Recognizes that sustainability is relevant to all aspects of school life including formal and hidden curricula, school leadership and management, as well as teacher development. Whole-school approaches encourage schools to practice what they preach. (Ferreira, Ryan, & Tilbury, 2006)
Nature-Based Learning	Nature-based learning (NBL) describes educational experiences that take place in the natural world, with nature being the subject of, and/or context for, learning. NBL includes informal (play), non-formal (community-based), and formal (school-based) learning. NBL also references situations in which nature serves as a complement to learning, such as the presence of indoor plants, access to green views while inside, and the occurrence of natural elements on school grounds and in the surrounding neighborhood. (Chawla, 2018; Jordan & Chawla, 2019)

Source: Kensler & Uline (2017, p. 184)

All these approaches underscore a focus on environmental issues and related content. Some extend beyond environmental concerns, to include the economic and social dimensions of life on our planet. Most stress the examination of values and seek subsequent changes in behavior. They urge learning within a local context but require the application of global perspectives. All approaches place students at the center, engaging them in active, place-based, problem-based, nature-based, and community-connected learning to encourage participation and critical thinking. In most cases, academic content is presented in a rigorous, interdisciplinary fashion, informing students' efforts to solve real-world problems (Kensler & Uline, 2017).

All of these approaches depend upon teachers building some foundational knowledge of sustainability concepts. In his recent book, entitled *Educating for Sustainability: Principles and Practices for Teachers*, Victor Nolet (2016) identifies a number of Big Ideas, in the manner of Wiggins and McTighe (2005), as over-arching constructs or themes that also inform a sustainability worldview (i.e., "a way of seeing and engaging in the world through the lens of sustainability", p. 10). By no means a comprehensive list of sustainability-related concepts, Nolet suggests these eight provide a minimal set that, once understood, will "assist teachers in prioritizing content, identifying learning progressions, developing long-term plans, anticipating students preconceptions, and guiding student learning" (Nolet, 2016, p. 68). Nolet's eight big ideas, including *equity and social justice, peace and collaboration, universal responsibility, health and resiliency, respect for limits, connecting with nature, local to global,* and *interconnectedness,* provide a point of departure for teacher learning.

Equity and justice stands at the center of a sustainability worldview. This big idea incorporates a number of related concepts, including social justice, economic justice, environmental justice, gender equity, food justice, climate equity, and intergenerational equity. Through the lens of equity and justice, we consider access to resources and opportunities, confront notions of privilege, and distinguish wants from needs.

Peace and collaboration, as a big idea, recognizes our fundamental human need for amity and security. It also underscores the environmental conditions that threaten peace. As we come to understand the interconnectedness of human and natural systems, we appreciate how peace and collaboration contribute to a healthy planet, while conflict between people and nations has deleterious effects on both the human and the natural world.

Universal responsibility holds us to account for the consequences of our decisions and actions. This big idea underscores our obligation to "promote the creation of a safe and just space for all forever" (p. 74). Universal responsibility requires we refrain from doing harm and *more*. Here we expect our students to realize their potential as active agents in creating a sustainable future.

Health and resiliency offers students "a context for investigating the characteristics of healthy, thriving systems" (p. 75). We consider questions about individual health and well-being, on the one hand, and large-scale concerns such as hunger and disease, on the other. A focus on resilience encourages an asset-based stance that underscores an individual's or a system's ability to change, to weather hardship or trauma, and to continue functioning and developing. This big idea also allows students to examine their own capacity to adapt to life events.

Respect for limits acknowledges the earth's finite capacity to support the survival of its inhabitants. This big idea calls for a transformation in the way we think about our relationships with one other, with the generations who will follow us, with other species. Students are called to question their own wants and needs as they examine the implications of consumption in relation to notions of fairness and justice.

Connecting with nature addresses the ways humans interact with the natural world. We explored this big idea in greater depth in Strategy 5 as we consider the critical role that contact with nature plays in children's well-being, development, and learning. As students learn in and from nature, they gain new respect for the elegance, balance, efficiency, and adaptability of natural systems. "Learning in and from nature also help learners integrate otherwise abstract theoretical concepts into a more active and personal understanding" (p. 77).

Local to global captures the interdependence of our social, economic, political, and natural systems worldwide. It further underscores how our local actions impact the global community; indeed, it reminds us that we are all members of this global community, for good or bad. We have returned repeatedly to this big idea across chapters in this book, demonstrating how education for sustainability engages students in the places they inhabit as a practical and profound means to connect them with the wider world.

Interconnectedness draws our attention to interwoven human and natural systems, i.e., social-ecological systems (Nolet, 2016). As a big idea, it leads students to engage in systems thinking, examining how individual elements of a system interact with one another (e.g., organisms within an ecosystem) or grappling with complex, large-scale environmental, social, and economic systems that extend globally, with reach and influence across nations and regions.

Together, these sustainability-related big ideas provide a "robust framework for exploration of content standards and application of critical thinking, problem solving, and systems thinking" required of students as they are assessed on the Common Core State Standards (CCSS) and the Next Generation Science Standards (NGSS) (Nolet, 2016, p. 102).

Nolet's (2016) book provides teachers a solid foundation in sustainability-related knowledge. He further recommends teachers explore exemplars of effectively implemented sustainability education that has resulted in positive student outcomes (Nolet, 2009), including, for example, *Ecological Literacy: Educating Our Children for a Sustainable World* (Stone & Barlow, 2005); *The Earth Charter In Action: Toward a Sustainable World* (Corcoran, Vilela, & Roerink, 2005); *Place-Based Education: Connecting Classrooms and Communities* (Sobel, 2005); and *Learning Gardens and Sustainability Education: Bringing Life to Schools and Schools to Life* (Williams & Brown, 2012).

Offering Teachers All Necessary Resources

Daily, teachers wrestle with "the exquisitely fragile relationship among knowing, valuing, and doing" (Nolet, 2009, p. 429). Many teachers question their individual and collective capacity to transform classroom practice in ways that increase students' engagement and spark their love for learning. There is good news for teachers, however, for sustainability, by its very nature, "creates an accessible path to help new [and seasoned] teachers link discipline and pedagogical content with real-world problems [in ways that] captures the contextual nature of deep learning" (Nolet, 2009, pp. 432, 429).

Still, teachers need continuous, strategic support to accomplish what, for many, amounts to a reinvention of their teaching practices, including curriculum, instruction, and learning context. As teachers venture out into the natural world with their students, they will need pathfinders to mark the way and carefully curated materials and supplies to guide and enrich the journey. Living-systems minded school leaders within Encinitas Union School District (EUSD) have restructured their curriculum development processes to better support this transformation in teaching practice. Rather than using district funds to purchase premade science curriculum, the district utilized these resources to reassign four classroom teachers as Teachers on Special Assignment (TOSA). Julie Burton, Coordinator for Innovation and Development at the EUSD Farm Lab, explained,

> Classroom teachers across the district applied for the Teacher on Special Assignment (TOSA) positions, both to write curriculum and support other teachers in classrooms. It's teachers teaching teachers. The TOSAs also have time to attend additional trainings themselves. They then pass on this learning as they model lessons for teachers.

EUSD's TOSAs were tasked with writing Next Generation Science Standards-based curriculum, which also integrates key sustainability-related concepts and skills. Consistent with EUSD's identification of *Environmental Stewardship* as one of the key Pillars of Distinction, guiding all district goals (see the Introduction of this book for further discussion of the Pillars), each unit foregrounds the NGSS instructional segments that best lend themselves to deep sustainability-related learning. In Figure 4.1, DREAM Campus teachers post important NGSS concepts as reinforcement when students visit their classrooms.

Instructional units for all grade levels are constructed in accordance with the 5E model of engage, explore, explain, elaborate, and evaluate, to avoid front-loading of information and, instead, allow students to develop an understanding through engagement and exploration. The entire curriculum is available to all teachers via the PowerSchool (www.powerschool.com/) digital platform, complete with a rich array of relevant resources, "more resources than any teacher could ever use," says Burton. She elaborates,

> Once the instructional segments are complete for a grade level, all the teachers for that grade level, across the entire district, come to one site. The TOSAs walk them through the units, and the teachers go hands-on with some of the experiences. They explore, feel, and utilize the curriculum in order to get really comfortable with it. So, there's no, "Here-you-go" handoff. It's always, "Let's come together, log in, and collaborate with one another."

Figure 4.1 NGSS cross cutting concepts are displayed in a DREAMS Campus classroom.

In addition to the online resources, the TOSAs developed corresponding supply lists for each unit. The school district then funded grade-level supply kits for every school. These kits contain all the materials needed for each instructional segment. Once again, rather than purchasing premade curriculum, EUSD paid their own teachers to write the new science curriculum so it might be truly place-based and aligned with local context, including the district's key pillar of *Environmental Stewardship*.

Dr. Baird underscored the ways district office roles were reinvented in order to provide teachers all necessary resources.

> District support exists to help teachers do their job, and that fact needs to be clearly communicated. Teachers don't exist to serve the district office. We're here to make sure they can do their job to the best of their capabilities and that they have the tools, the resources, the knowledge, and everything else to make it happen.

Next on the school district's agenda, TOSAs will begin the development of mathematical problem sets and resources in accordance with *Cognitively Guided Instruction* (CGI). CGI is a student-centered approach to teaching math, building on students' inherent number sense and problem-solving skills. "CGI is a way of listening to students, asking smart questions, and engaging with their thinking—all with the goal of uncovering and expanding every student's mathematical understanding" (www.heinemann.com/cgimath/). Dr. Amy Illingworth explained how teachers will engage in a three-year process of professional development.

> You cannot buy CGI curriculum. It doesn't exist. The goal is that teachers write their own problems, differentiating based on the students' levels. We will facilitate collaborative problem writing and provide teachers with problem sets and resources, with the goal of phasing teachers away from the traditional textbooks.

EUSD's proposed implementation of CGI provides yet another example of how school and district leaders provide the necessary resources to support locally developed, problem-based transformations in teaching practice across all district classrooms and schoolyards.

Invent Systems and Structures to Support Preparation of Teachers to Teach in Nature

Nolet (2016) reminds us that no "one discipline claims ownership [of sustainability-focused learning], all disciplines share responsibility, . . . includ[ing] content from quantitative reasoning/math, inquiry/natural sciences, creativity/the arts, critical thinking and policy analysis, social studies, communication and media, literacy/language arts, spatial reasoning/ geography" (p. 8–9). Such an integrated view of curriculum and instruction pushes living systems-minded school leaders to invent new systems and structures to facilitate this cross-disciplinary work. Teachers must plan, organize, research, and experiment together in a collaborative fashion.

Living systems-minded school leaders cultivate collaborative professional learning communities within which teachers reflect deeply and critically on their own teaching practice, on the content and context for learning, and on the experiences and backgrounds of the students in their classrooms. Living systems-minded school leaders play a vital role in providing and protecting sufficient time for teachers to meet in professional community with one another, guiding the vision and related professional development goals that will be pursued during this time and providing necessary resources to support the ongoing inquiry that is central to these collaborations (Louis, Marks, & Kruse, 1996; Mullen & Hutinger, 2008; Olivier & Hipp, 2006).

In EUSD, teachers meet together in grade-level teams weekly for a half-day of collaboration, during which they plan units of instruction. EUSD schools referred to these collaborative planning times as "Wheel Time" or "Team Time." While a given grade-level team collaborates, their students experience four rotations of special classes ("The Wheel") taught by Site Enrichment Teachers. EUSD's Site Enrichment Teachers are paid a daily rate. They are credentialed teachers who work part-time and, therefore, do not earn a full-time teacher salary. These positions are partially funded by the district, with the remainder paid for by PTA and the Encinitas Education Foundation. Funds are raised specifically for these positions so that the teachers have adequate planning time.

Special classes vary by school, based on particular needs. Common options include garden and nutrition, music, art, and yoga exercise. In some schools, students attend an extra science lab. All students participate in yoga class, as seen in Figure 4.2. Julie Burton explains,

> It's not prep time. It's collaboration time, because the school district is committed to equity across each grade level. It doesn't matter which teacher your child receives. Students are going to get a solid, comprehensive experience. We don't have just one teacher in a grade level engaging students in project-based learning. They're all doing it, because they have regular, guaranteed time to plan together while their students go through the Wheel.

Wheel Time provides a structure within which teachers work together to explore district-provided curriculum resources as a team. They begin by investigating various learning and performance data in order to determine their students' background knowledge and current learning needs. They then modify and/or design units accordingly. One EUSD principal explained why such regular planning time was essential to realizing the district's goals related to equity and excellence.

> It (Wheel Time) makes a huge difference. As a principal, I can truthfully say, what's happening in one classroom is happening in all the other classrooms. It might look differently, because of each teacher's individual personality and style, but they're all delivering the very same unit. They're utilizing the same assessments. For example, the kindergarten grade-level team delivers a weather unit. That's part of the standards, so that's what we do. But Wheel Time allows grade-level teams to move beyond the standard. Students are learning about how forecasting weather relies upon predicting patterns, and they then look at patterns in math and patterns in poetry. Teachers are able to provide a great, holistic vision. This deeper learning requires a ton of planning. The Wheel allows for that.

As living systems-minded school leaders investigate possible structures and systems to support the implementation of sustainability-focused, nature-based teaching practices, they are wise to direct their inquiry toward "both the classroom conditions that students experience directly and the wider organizational conditions that enable, stimulate, and support these conditions" (Leithwood & Sun, 2012, p. 413). In order to successfully shift learning and teaching out beyond the classroom walls, leaders must invent systems and structures to support changes in instructional and curricular norms of practice. ESUD's locally developed curricula, in concert with regularly scheduled

Figure 4.2 Students participated in yoga class during Wheel Time.

teacher collaboration times, provide necessary supports as EUSD teachers embark upon this challenging, yet creative and motivating, work.

 ## Model Effective Nature-Based Teaching Practices

Adopting a new practice often requires individual teachers to relinquish a more familiar or comfortable practice. As teachers are encouraged to utilize the natural world as their content and their classroom, direction and feedback from teachers, who are experienced in and comfortable with such nature-based practices, helps to quell fears. Kuo, Browning, and Penner (2018) conducted an experiment designed to test the validity of this fear teachers have about taking students outside. Many teachers believe that time in nature disrupts students' ability to concentrate. In fact, evidence from Kuo et al.'s (2018) experiment suggests students' concentration actually improved with time spent in nature.

Teachers no doubt span the career continuum, with some seasoned professionals ready to jump feet-first into interdisciplinary, learner-centered, problem- and place-based instructional practices. Beginning teachers may be less confident about their ability to facilitate and manage the uncertainty and ambiguity that comes with challenging the traditional architecture of instruction. All teachers must learn to guide student inquiry, let questions emerge, and be more comfortable with the complexity and uncertainty that results (Lowenstein et al., 2010).

> As teachers engage in challenging learning both in and out of the classroom, teachers require a sense of their own efficacy. As teachers work with students over time and see that their students are capable of engaging in the kinds of deep thinking and reasoning they never thought possible, teachers' own sense of efficacy increases.
>
> (Woolfolk Hoy, Davis, & Pape, 2006, p. 734)

As living systems-minded school leaders seek to develop their teachers' sense of efficacy for teaching in nature, they incorporate multiple opportunities for structured professional development. Much of this can and should be self-initiated and directed, utilizing locally sourced expertise of their own teachers to teach one another.

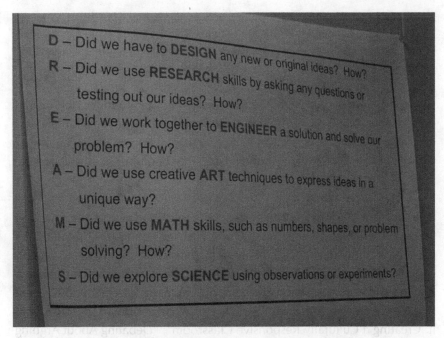

Figure 4.3 Discussion prompts are displayed in a DREAMS Campus classroom.

EUSD's DREAMS (Design, Research, Engineering, Math, and Science) Campus provides a powerful vehicle for modeling "the passion-based, project-based, real-world learning experiences" district leaders hope will propagate throughout the district. DREAMS Campus teachers include one TOSA and two enrichment teachers who are paid on an hourly rate. These teachers model project-based learning strategies, as indicated by the widely utilized discussion prompts shown in Figure 4.3. Burton describes the school district's vision for utilizing this valuable learning resource.

The DREAMS Campus takes everything the district is attempting to accomplish and models it for teachers. The DREAMS Campus provides a physical place where teachers see the shift. "Come see it in action. Come learn along with your kids and see how engaging this is. See how you can integrate the content areas and teach the conceptual understandings that will matter in 10 years. Observe how to teach students that they have a voice and that their actions make a difference that can be positive or negative." It's not a formal professional development experience, but [DREAMS Campus instructors] model the delivery of interdisciplinary units and teachers experience the active learning along with their kids. You'll hear teachers say, "Whoa, isn't it a little chaotic in here?" And, the DREAMS Campus teacher will respond, "But, listen. What are they talking about? What are you noticing?" And teachers settle into the idea, "Oh, this is actually purposeful. The kids are engaged. They're not distracted." Teachers can observe that we have almost no behavior issues out here.

In addition to DREAMS Campus modeling opportunities, the district TOSAs, mentioned earlier in this chapter, also work directly with teachers who may be timid. TOSAs travel to individual classrooms across the district to guide and model, rotating among different school sites so all teachers get to know and trust them.

EUSD also holds four district-wide, half-day release professional development sessions each year. Teachers fully lead one of the four First Friday Professional Development Days. Each session lasts 45 minutes. All sessions include strategies for differentiating instruction and tools to support students learning English, students with disabilities, and advanced learners. Materials from every session are made available on PowerSchool. A recent Friday schedule included topics such as "What's Culture Got to do with It: Creating a Culturally Responsive Classroom"; "Debating About Arguing on Purpose: Growing Listening and Speaking Skills with Purpose"; "Real Science, Real Change, Real Fun: NGSS-Aligned Projects, Stewardship, and Community Action"; "Understanding Our Children with Autism"; and "Share a Lesson Goes Digital: Grades 4–6!" Here again, teachers teach teachers, and teachers choose which learning experiences will best suit their needs.

Introduce Teachers to Their School Facility as a Three-Dimensional Textbook

As teachers are better apprised of the design intentions for their school building, they may be motivated to collaborate more, experiment more, and venture out across the landscape beyond the classroom walls. Learning the way building and grounds work, and why, constitutes an important focus for teacher professional development.

Building occupant training is most effectively provided through a commissioning process, and, in the case of new buildings, should span from pre-design through occupancy, and even beyond (Lackney, 2005). In one of its earliest uses, commissioning was a process utilized for testing

U.S. naval ships to ensure their reliability and quality prior to deployment. Commissioning was later integrated into the language of facility professionals to describe a process for validating a building's performance in relation to operational requirements and design intent (Lackney, 2005). Lackney and colleagues coined and trademarked the term *educational commissioning* to refer to a specific context and process "through which teachers, students, and even parents and community partners are educated as to the design intent of a newly constructed school facility" (Lackney, 2005, p. 1). While similar to previous conceptions of the term, this iteration moves beyond the original idea of simply evaluating quality and fit. By involving, educating, and training occupants and stakeholders throughout the design process and through move-in and beyond, we increase the likelihood that the building will support the specific learning and teaching needs of those who inhabit it day to day. This learning also results in a higher probability of successful and effective future designs (see Kensler & Uline, 2017 for further discussion).

Dr. Baird voiced his frustrations regarding the many missed opportunities that result from inadequate facilities-related orientation and training.

> Districts call themselves green, because they install solar panels on a building, but nobody inside the building even knows the panels are there. Every physical change is a learning opportunity. Every time you do anything green, whether it's with the custodians, the maintenance staff, or the food service staff, it's a potential learning opportunity. When you install solar panels on buildings, you now have a built-in math project. You can track how much energy you're using. You can track how much energy you're saving. You can compare building to building. We often use water use as a learning opportunity. We wrote a grant through the State of California that funded best water management practices. Students designed bioswales and worked with contractors to create them.

When school building occupants are more informed about possibilities and more involved in defining parameters, acclimation to sustainable features and technologies is assured, and users are better able to envision how these can be leveraged to change the way learning takes place. As educators have opportunity to learn green buildings, "they begin to use the world of the physical environment as a teaching tool to help students understand the underlying laws and principles that govern our complex, precious universe" (Taylor, 2009, p. 3). Revisit Strategy 3: MAINTAIN AND

OPERATE to find suggestions for utilizing the school facility as a three-dimensional textbook.

Conclusion

Nature-based, sustainability-focused learning provides living systems-minded school leaders and their teachers an inspiring and motivating approach to improving instructional effectiveness at their schools. Nature-based, sustainability-focused learning invites teachers to experiment with increased levels of integration across content areas. Nature-based, sustainability-focused learning offers a ready context for learner-centered, problem- and place-based learning experiences that spurs students to stretch for higher, more complex levels of subject matter thinking. Nature-based, sustainability-focused learning inherently sows the very knowledge and skills necessary for college and careers, challenging students from the primary grades on to engage in rigorous, integrated, and context-embedded curricula. As Nolet (2016) reminds us, "It would make no sense for something called education for sustainability to interfere with helping all learners develop a strong foundation in basic academic skills and knowledge . . . teaching for sustainability should enhance and augment teachers' day-to-day professional practices" (p. 10), at the same time it renews their excitement and vigor for teaching.

 Leadership Design Challenges

1. **Develop a plan for instituting regular teacher collaboration time at your school**. Prepare a detailed description of the structures and processes that will guarantee your teachers have sufficient time to meet, engage, and learn together on a regular basis. Include necessary strategies for protecting this time and ensuring it remains focused and productive.

2. **Identify a small cadre of teacher(s) who are comfortable teaching in nature. Design a process that sets them loose to model for their peers**. This will take some creative planning, but the dividends it pays will be well worth the effort. The time and resources you invest in designing this peer-modeling opportunity for your teachers will demonstrate

your commitment to nature-based, sustainability-focused learning and your openness to risk taking as a necessary means to improve instructional effectiveness.

 Learning From Living Systems-Minded Trailblazers

Out Teach is a nonprofit organization whose mission is to "equip teachers with the power of experiential learning outdoors to unlock student performance" (www.out-teach.org/). They serve schools and school districts across the United States, including Washington, DC, Virginia, Maryland, Georgia, North Carolina, and Texas. With an array of funding partners, Out Teach is rapidly expanding their geographic reach. Each time they partner with a school, they bring together funding partners and volunteers to design and build outdoor classrooms. With more engaging outdoor spaces, they then lead and facilitate teacher learning so that teachers are equipped to better care for and utilize these spaces for student learning across science, math, and literacy. They prioritize work with schools facing most challenging economic circumstances, thus making high-quality outdoor learning available to all students. In fact, their vision is "to ensure that all students, no matter their resources, have access to an engaging, hands-on education that transforms their lives" (website listed above).

Out Teach has realized impressive results. Participating teachers report higher levels of job satisfaction and effectiveness. Nearly all participating teachers (94%) report increasing student engagement, due to using lessons they learned in their Out Teach professional development. These teachers also report being better prepared to support their students' academic performance. District-level, student-performance data suggest this is so, with partner-school proficiency rates on standards-based assessments increasing by 12–15% overall. Statistical analysis suggests that Out Teach programming is responsible for one third of these gains.

The Out Teach website includes testimonials, such as this one from the Principal of Ketcham Elementary School in DC Public

Schools (DCPS), Maisha Riddlesprigger. Riddlesprigger has served as principal in the Anacostia neighborhood of Washington, DC since 2013, earning recognition as DCPS Principal of the year in 2019. She shared the following thoughts on the Out Teach website,

> This outdoor classroom increases the J-factor at our school, and by that, I mean Joy—Joy of learning and simple joy in these kids' lives every day. They're going to look forward to coming to school, because of our partnership with Out Teach. I had a student so excited about a lesson where they'd been measuring, she came running through the halls to come get me and share her experience with me. She had a tape measure in one hand and an onion she'd grown in the other. She showed me exactly how it was done. She was so proud, so thrilled. That's what school means to her now. Excitement and pride. The learning garden has been a great way to engage students in social emotional learning. The outdoor learning lab is great for kids with I.E.P.s and Special Needs. They're able to stay engaged in the lessons with the other kids. It's easier for teachers to make sure they're not being marginalized.

Out Teach is cultivating teachers' capacity to teach outdoors across the United States. These teachers are engaging 57,000 students each year in joyful, outdoor learning across STEM, 21st-century skills, social emotional learning, and health and wellness.

References

Bauermeister, M. L., & Diefenbacher, L. H. (2015). Beyond recycling: Guiding preservice teachers to understand and incorporate the deeper principles of sustainability. *Childhood Education*, *91*(5), 325–331. doi:1080/00094056.2015.1090843

Chawla, L. (2018). The American Association for Teaching & Curriculum (AATC) keynote address: Nature-based learning for student achievement and ecological citizenship. *Curriculum and Teaching Dialogue*, *20*(1 & 2), xxv–xxxix.

Corcoran, P. B., Vilela, M., & Roerink, A. (2005). *The earth charter in action: Toward a sustainable world*. Amsterdam: KIT.

The Earth Charter International. (2015). *Earth charter: Values and principles for a sustainable future*. Retrieved from www.earthcharterinaction.org

Ferreira, J., Ryan, L., & Tilbury, D. (2006). *Whole-school approaches to sustainability: A review of models for professional development in pre-service teacher education*. Canberra: Australian Government Department of the Environment and Heritage and the Australian Research Institute in Education for Sustainability (ARIES).

Jordan, C., & Chawla, L. (2019). A coordinated research agenda for nature-based learning. *Frontiers in Psychology, 10,* 766. doi:10.3389/fpsyg.2019.00766

Kensler, L. A. W., & Uline, C. L. (2017). *Leadership for green schools: Sustainability for our children, our communities, and our planet*. New York: Routledge/Taylor and Francis Group.

Kuo, M., Browning, M. H. E. M., & Penner, M. L. (2018). Do lessons in nature boost subsequent classroom engagement? Refueling students in flight. *Frontiers in Psychology, 8,* 1–14. https://doi.org/10.3389/fpsyg.2017.02253

Lackney, J. (2005). Educating educators to optimize their school facility for teaching and learning. *Design Share*. Retrieved from www.designshare.com/index.php/articles/educational-commissioning/

Leithwood, K., & Sun, J. (2012). The nature and effects of transformational school leadership: A meta-analytic review of unpublished research. *Educational Administration Quarterly, 48,* 387–423.

Louis, K. S., Marks, H. M., & Kruse, S. (1996). Teachers' professional community in restructuring schools. *American Educational Research Journal, 33,* 757–798.

Lowenstein, E., Martusewicz, R., & Voelker, L. (2010). Developing teachers' capacity for ecojustice education and community-based learning. *Teacher Education, 37*(4), 99–118.

McClam, S. & Diefenbacher, L. (2015). Over the fence: Learning about education for sustainability with new tools and conversation. *Journal of Education for Sustainable Development, 9*(2), 126–136.

Mullen, C. A., & Hutinger, J. L. (2008). The principal's role in fostering collaborative learning communities through faculty study group development. *Theory into Practice, 47,* 276–285.

Nolet, V. (2009). Preparing sustainability-literate teachers. *Teachers College Record, 111*(2), 409–442.

Nolet, V. (2016). *Educating for sustainability: Principles and practices for teachers.* New York: Routledge.

Olivier, D. F., & Hipp, K. K. (2006). Leadership capacity and collective efficacy: Interacting to sustain student learning in a professional learning community. *Journal of School Leadership, 16,* 505–519.

Riordan, M., & Klein, E. J. (2010). Environmental education in action: How expeditionary learning schools support classroom teachers in tackling issues of sustainability. *Teacher Education Quarterly, 37*(4), 119–137.

Sobel, D. (2005). *Place-based education: Connecting classrooms and communities.* Great Barrington, MA: The Orion Society.

Stone, M. K., & Barlow, Z. (Eds.). (2005). *Ecological literacy: Educating our children for a sustainable world.* San Francisco: Sierra Club Books.

Taylor, A. (2009). *Linking architecture and education: Sustainable design of learning environments.* Albuquerque: University of New Mexico Press.

Thomas, G. (2005). Facilitation in education for the environment. *Australian Journal for Environmental Education, 21,* 107–116.

Wals, A. E. J. (2009). *Review of contexts and structures for education for sustainable development 2009: Learning for a sustainable world.* Paris, France: UNESCO.

Wiggins, G., & McTighe, J. (2005). *Understanding by design* (2nd ed.). Alexandria, VA: ASCD.

Williams, D. R., & Brown, J. D. (2012). *Gardens and sustainability education: Bringing life to schools and schools to life.* New York: Routledge.

Woolfolk Hoy, A., Davis, H., & Pape, S. (2006). Teachers' knowledge, beliefs, and thinking. In P. A. Alexander & P. H. Winne (Eds.), *Handbook of educational psychology* (2nd ed., pp. 715–737). Mahwah, NJ: Lawrence Erlbaum.

STRATEGY

5

Learn
Invite Students and Teachers to Learn in Nature

Living systems-minded school leaders intentionally disrupt the traditional architecture of instruction. They open doors and invite learning to spill out beyond classroom and school walls. Here students encounter the natural world. When children are allowed to experience nature, with all its colors, shapes, textures, sounds, and smells, they are apt to respond with excitement and curiosity. Educators are wise to tap these responses as motivators and sources of improved attention to learning (Bølling, Otte, Elsborg, Nielsen, & Bentsen, 2018; Kuo, Browning, & Penner, 2018). In fact, time in nature presents a valuable opportunity for students to learn complex concepts and develop important academic skills (Camassoa & Jagannathan, 2018). As well, time in nature promotes students' overall health and well-being, including specific aspects of social/emotional, physical, and cognitive well-being, all of which are foundational to students' engagement in learning (Kuo, Barnes & Jordan, 2019).

As leaders and teachers seek to meet the current challenge of educating students amidst the COVID-19 pandemic, history provides a promising solution to necessary social distancing and clean airflow. During the tuberculosis outbreak that ravaged U.S. cities in the early 20th century, 65 fresh-air schools, comprised of classrooms with windows on four sides or where classes were "simply held outside", opened across the country (Korr, 2016, p. 75). In these schools, students who had been exposed to TB improved rapidly and avoided exposure to other contagious diseases, including the common cold (Korr, 2016). A July 17, 2020 *New York Times* article described Mayor Bill de Blasio's plan to allow public, private, and charter schools to hold in-person classes outdoors as schools reopen in the city (Bellafante, 2020). It should be noted that systemic inequities across the district present a different set of challenges for schools in low-income neighborhoods, where green space and safe streets are limited. According to de Blasio,

> It really depends on the circumstances of each school, but one thing we know for sure is we're going to focus on the 27 neighborhoods hardest hit by COVID-19. We are going to prioritize making sure that they get options for outdoor space.
>
> (Bellafante, 2020)

This strategy outlines various steps living systems-minded school leaders take as they (1) prioritize regular access to nature and the outdoors; (2) promote learning in nature as a viable means to advance students' cognitive well-being and academic growth; (3) focus nature-based learning experiences in ways that contribute to students' social and emotional growth; and (5) design authentic learning experiences in nature to consciously develop students' capacity as local and global citizens.

Prioritize Regular Access to Nature and the Outdoors

A significant body of research demonstrates a positive relationship between access to nature and the well-being of children and adults (Gill, 2014; Kuo &

Faber Taylor, 2004; Louv, 2008; Noddings, 2013), the healing effects of views of nature (Li & Sullivan, 2016; Ulrich, 1984), and nature's positive influence on mental health (Zarghami & Fatourechi, 2015). Further, outdoor activities have been shown to encourage more creativity than those in classrooms (Lindholm, 1995). Recent reviews of an expanding literature present strong evidence that experiences in nature enhance students' educational and developmental outcomes (Jordan & Chawla, 2019), including academic learning, personal development, and environmental stewardship (Kuo et al., 2019). According to Kuo et al. (2019),

> The evidence here is particularly strong, including experimental evidence; evidence across a wide range of samples and instructional approaches; outcomes such as standardized test scores and graduation rates; and evidence for specific explanatory mechanisms and active ingredients. Nature may promote learning by improving learners' attention, levels of stress, self-discipline, interest and enjoyment in learning, and physical activity and fitness. Nature also appears to provide a calmer, quieter, safer context for learning.
>
> (Kuo et al., 2019, p. 1)

For these reasons, living systems-minded leaders start by identifying and/or creating places where students and teachers can regularly access nature and the outdoors. Toffler and Louv (2020) remind us that, in most communities, public school districts are one of the top local landholders, "pointing to a great and often underutilized resource that can be reimagined to support student achievement and community well-being, as well as mitigate the effects of climate change" (p. 69–70). A school site can provide learning landscapes (Taylor, 2009) with interconnected features including natural (climate, plants, animals, soil and rocks, wetlands), built (play structures, bermed earth, pathways, sports venues, seating), multisensory (texture, patterns, colors, patterns, sounds, smells), cultural (gathering spaces, local materials, student and public art, indigenous design), agricultural (gardens, land management areas, orchards), and outdoor classroom (weather stations, solar and wind energy stations, trails, greenhouses, water harvesting systems) elements (Taylor, 2009). In combination, these features provide diverse settings for a multitude of learning activities (Gelfand, 2010).

A number of Encinitas Union School district (EUSD) school campuses maximize the use of learning gardens on their school sites (see Figure 0.1. Encinitas Union School District Green Initiatives, beginning on p. 22). For example, the district's oldest campus has established their overall theme as "A School Within a Garden", with eight different gardens planted across the site, including one that relies on hydroponics. The main garden serves as an instructional garden. The principal of another EUSD campus describes the intricacies of an entire learning garden system on her campus.

We created our learning garden a long time ago, and now it is fully parent-run. The students engage in garden science. We conduct experiments in the garden beds and with the harvested food to prepare meals in the culinary lab (See Figure 5.1). We utilize vermicomposting [a composting process using various species of worms], with the kids collecting the necessary vegetable waste at lunch. You'll also see the gray water sinks with buckets underneath. We recycle

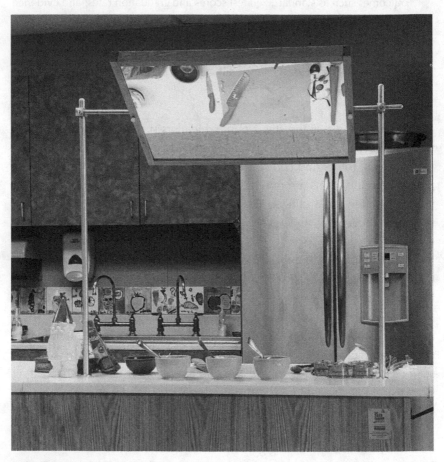

Figure 5.1 Students experiment in this Culinary Lab at one EUSD Elementary School.

and reuse that water. Along with vegetables crops, we grow fruit trees. At the middle of harvesting, we hold a school-wide community garden day.

As students venture outside to tend their school's learning garden(s), they have opportunity to conduct experiments, solve mathematical problems, prepare written explanations of their actions, and/or make literary and art works to capture the mystery and beauty of the natural world.

As mentioned in earlier sections of this book, all 5,400 EUSD students visit EUSD's Farm Lab and DREAMS (Design, Research, Engineering, Math, and Science) Campus, a ten-acre innovative indoor and outdoor educational campus (See Figure 5.2).

Farm Lab is the first in the nation to become a school-district-owned, certified-organic crop production farm, supplying the district's own school lunch program (See Figure 5.3). Here students examine the local food environment, learning how foods might be sourced locally, organically, and with minimal processing.

Each grade level also accesses the DREAM Campus and Farm Lab for up to a full week of learner-centered, problem- and place-based learning experiences (See Figure 5.4). As mentioned in Strategy 4 TEACH, DREAMS Campus teachers, including one TOSA and two enrichment teachers, lead students and their teachers through standards-based, interdisciplinary units of instruction.

Figure 5.2 Farm Lab DREAMS Campus boasts indoor/outdoor classroom facilities.

Figure 5.3 Farm Lab garden beds provide food for the district school lunch program.

For example, in accordance with Next Generation Science Standards (NGSS), kindergarten students learn about the growth cycle of plants, from seed to compost (See Figures 5.5 and 5.6). Second-grade students investigate pollination. Fourth-grade students engage in environmental stewardship through the study of renewable energy. They learn about the impact of fossil fuels, investigate the benefits of various renewable energy sources, and design a wind turbine, attempting to harness the wind efficiently enough to light an LED bulb.

As stated earlier, EUSD's living systems-minded school leaders turn their school campuses inside out and outside in, blurring the lines between built and natural environments. In so doing, they invite students to observe how these spaces are necessarily connected and interdependent, opening the door (literally and figuratively) for ecologically sensitive learning to take place.

Promote Learning in Nature as a Viable Means to Advance Students' Cognitive Well-Being and Academic Growth

Recently, nature-based learning (NBL) has emerged as a term used to describe educational experiences that take place in the natural world, with nature being the subject of, and/or context for, learning (Chawla,

Figure 5.4 Farm Lab DREAMS signage announces the deep learning that takes place on campus.

2018; Jordan & Chawla, 2019). NBL includes informal (play), non-formal (community-based), and formal (school-based) learning. NBL also references situations in which nature serves as a complement to learning, such as the presence of indoor plants, access to green views while inside, and the occurrence of natural elements on school grounds and in the surrounding neighborhood (Chawla, 2018, p. xxvii). "What is common across these different facets of NBL is that nature is accessible to children. At a minimum, they can see it, and in other cases they can experience it through all senses (Chawla, 2018, p. xxvii). In this book, when we talk about *taking students outside* and *bringing nature inside*, we acknowledge all the possibilities encompassed within these understandings

Figure 5.5 Kindergarteners study plant life as they grow pumpkins at farm Lab.

of NBL. We, too, are interested to acknowledge "the impact of natural surroundings on learning in general" (Chawla, 2018, p. xxvii), in order to advance learning in nature as a legitimate approach to learning and teaching in schools.

Academic performance depends in large part on students remaining focused and engaged in their learning, and emerging research consistently

Figure 5.6 Hay bales provide seating for group discussions.

demonstrates that time in nature improves students' capacity to focus (Chawla, 2015; Kuo et al., 2018). A recent study, found that 9- to 11-year-old students' attention to task was best after time in woods, the context that students also reported as most restorative, as compared to time in the classroom or on the playground (Berto, Pasini, & Barbiero; 2015). Fourth-grade students, in schools with improved schoolyards, performed better on standardized testing, even after controlling for demographic characteristics, such as income and race (Lopez, Campbell, & Jennings, 2008).

Evidence suggests that nature-based learning may provide opportunities to address achievement gaps between student groups (Emekauwa, 2004). Three Louisiana elementary schools, where 80% of students were African American and 85% received free or reduced lunches, adopted curricula focused on the study of local natural resources. At the beginning of the first year of implementation, students in these schools scored 10.7% lower in math and 15.6% lower in social studies, when compared with state averages. By the end of the third year, these gaps were reduced to 0.2% and 7.5%, respectively.

A natural science and environmental education program, designed to help elementary students from low socio-economic backgrounds increase their knowledge of science and strengthen their overall academic performance, utilized specially designed naturescapes to facilitated active, hands-on learning (Camassoa & Jagannathan, 2018). Within these naturescapes, students learned about plant and animal life. An organic garden allowed for experiments in fruit and vegetable cultivation, pest control, hybridization, and nutrition. A water feature extended learning to fish, amphibian, and aquatic plant life. Students identified and classified species and observed complex processes such as metamorphosis and pollination. The program rested on a commitment to involve parents in their child's math and science education and included supplemental after-school and summer instruction, as well as math, language arts, and science tutoring. A four-year evaluation, structured according to an experimental design, revealed students within the program consistently outperformed students in a study control group in both mathematics and science, with the differences in science reaching statistical significance. Program students, who regularly experienced natural and environmental science lessons in situ, also had higher rates of attendance. Camassoa and Jagannathan (2018) observed that "[t]he academic performance of students, so highly valued by school administrators and parents, would appear to benefit within this type of environment, even as the students' appreciation for the wonders of nature flourishes" (pp. 274–275).

EUSD Assistant Superintendent Illingworth, described how EUSD's approach to providing nature-based learning helps them to meet important equity-related goals.

When you think about the different demographic groups of students we serve (for example, we have English learners at 80% of our schools), every student in this district has access to amazing learning experiences that don't happen in other districts. Every one of our students gets to visit the Farm Lab. They put their hands in a garden, produce something, and harvest something. And, every student engages in researching, speaking, listening, and, ultimately, presenting. They see the results of all this work.

Dr. Baird explained how teachers and leaders in EUSD understood learner engagement as fueled by *purpose, passion, power,* and *play.*

We talk about, "How do you engage learners?" We maintain you accomplish this through *purpose, passion, power,* and *play. Purpose* recognizes there must be meaning in the work. *Passion* involves deciding what you care about. If you care about it, you're going to be motivated and more likely to succeed. Then, *power* is giving the learner some control over time, team, task, and technique, those kinds of things. And, *play*—ultimately it should be fun. You should see kids smiling and enjoying the work that they're doing. So, environmental stewardship brings all these—meaning, caring about the outcomes, the potential for power, and the opportunity to play—automatically.

These four essential ingredients converge within meaningful, action-oriented learning in nature. Dr. Baird underscored the district's ongoing efforts to promote this nature-based learning as a highly effective means to facilitate students' academic growth.

I've had parents say, "You're experimenting on our kids." And, yes, in a way, educators have always been experimenting with students as they employ some treatment based upon a given hypothesis. The trouble is, we've been doing the same experiment for 50 years or more, and it's not working for too many students. So, let's try a new experiment and do something that matters to students. Let's do something that is meaningful. Let's engage kids. And, when we look at the data, our students are among the top scorers in this county and in this state.

Over time, these achievement data, combined with qualitative data (see a more detailed discussion of these data in the Introduction of this book) on students' enthusiastic engagement in learning, provided compelling evidence that the experiment was working (See Figure 5.7).

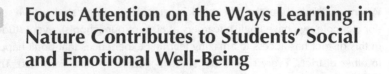

Focus Attention on the Ways Learning in Nature Contributes to Students' Social and Emotional Well-Being

Time in nature has the potential to be restorative, to reduce the effects of stress and nurture emotional well-being. Studies have shown that children across income levels, who spend time in nature, experience lower levels of stress, higher levels of self-worth, and lower rates of depression than those who do not (Chawla, 2015; Mustapa, Maliki, & Hamzah, 2015). When students were assessed before and after the greening of an Austrian middle school site, students reported significantly greater feelings of well-being after the redesign of their schoolyard (Kelz, Evans, & Röderer, 2015). In addition, students at the intervention school experienced reduced levels of blood pressure when compared with students at two control schools where schoolyards were not redesigned to include green elements (Kelz et al., 2015).

Figure 5.7 Students observe the transformation from tiny seedlings to delicious vegetables.

School gardens have been shown to provide additional spaces where children developing their social skills as they collaborate with one another to plant, tend, and harvest crops (Williams & Brown, 2012; Williams & Dixon, 2013). Children of all ages report feeling free, relaxed, and calm when in nature. In addition, adolescents who have had positive experiences in nature as young children are more likely to report that nature buffers their stress in adolescence (Chawla, 2015). To the degree that schools introduce children to nature at an early age, they increase the likelihood that older students will value and benefit from nature. A principal of one EUSD elementary schools described it this way:

> Just giving them [the students] time to be in the fresh air and outside those four walls (pointing to the school building), really opens it up. Right now, it's lunchtime, but after they finish eating, kids will be interacting at the log benches, playing on the ground, working in the garden—just really hands-on. The kids are watering the plants together. They're taking ownership of their school.

A principal of a Title 1 school in EUSD described how time outside benefited her English learners, in particular. She explained, "Twenty-five percent of our students are English learners. Getting students outdoors, and fully engaged in what they're doing, seems to make them feel comfortable having conversations and using their language."

Research confirms this principal's anecdotal observations, demonstrating that natural surroundings do indeed "provide a calmer, quieter, safer context for learning; a warmer, more cooperative context for learning" (Kuo et al., 2019, p. 1).

Design Authentic Learning Experiences in Nature to Consciously Develop Students' Capacity as Local and Global Citizens

Local, place-based, "school-in-community" learning experiences reaffirm that human life, culture, and society are firmly embedded within the natural world (Selby, 2000; Sobel, 2005). Place-based learning has long been a part of environmental education, aimed at deepening children's connection to nature, with investigations of natural ecosystems such as local creeks, ponds, forests, meadows, and coastal areas that are within a short walk from classrooms (Krapfel, 1999). This place-based learning provides important opportunities for students to understand the complex interdependencies at play between human and natural systems, motivating them to appreciate and care for these abundant places they call home, and by extension, to preserve the homes of other human and nonhuman inhabitants across the planet (Sobel, 2008).

Place-based learning opportunities might start small within a particular classroom, as teachers learn their way into this fundamentally different way of teaching and learning. Eventually, they may go school-wide, with place-based and/or thematic learning as the primary way learning happens for everyone (Meier, Knoester, & D'Andrea, 2015). A particular example of such pervasive place-based learning derives from studies of high-performing urban schools where educators demonstrated multiple evidences of success (e.g., state assessment scores, graduation rates, attendance rates, discipline data, English-acquisition data, course-taking patterns) for their students (Johnson, Uline, & Perez, 2019). These remarkable elementary, middle, and high schools (1) served predominantly low-income students; (2) did not use selective admissions criteria; and (3) achieved outstanding results for every racial/ethnic group served. In many of the high-performing urban schools studied, place/problem/project-centered learning dominated approaches to teaching science and mathematics. For example, at Pershing Elementary in Dallas, Texas, students learned in and from nature, growing and harvested crops, observing the butterfly cycle, harvesting rainwater, composting and fertilizing planted beds, observing and studying a local fish pond, and learning about xeriscaping (Johnson et al., 2019).

Place-based learning started small at the schools in Encinitas, where every school eventually had at least one garden within which teachers planned standards-based lessons. As teachers grew increasingly comfortable bringing students outside, one school went big. A principal described a kindergarten-level project that grew out of a field trip to a local lagoon.

Students took a trip to the San Elijo Lagoon. While they were at the lagoon, they learned that a lot of the marine life was dying, because they were digesting plastic straws and eating trash on the beach. The big thing that resonated with the kindergartners was the plastic straws. They learned that, to the sea birds, the straws look like worms, and so the birds eat the straws, and end up dying. They showed the kindergartners the inside of a stomach of one of the birds that had died, which really was a shocker. When they came back to school, they started writing me letters and sending me pictures.

"We can't have straws at lunch, and this is why."

I said to them, "You know, there's not much I can do about it at the school level. You really want to talk to Dr. Baird, because he is really in charge of the whole school."

Then, they asked, "Well, who's in charge of him?" and I said, "The school board is in charge of him."

They wrote letters, made posters, and gave a presentation to the school board about why we need to ban straws at lunch. The school board took action, and so that following school year, when the kindergartners were first graders, the school district took away the straws at lunch. We no longer offer straws at lunchtime. When the students heard this actually happening, they couldn't believe somebody had listened to them and, because of their action, the school district was changing their policy. Another school then took it a step further, and now the students are working to eliminate plastic spoons at lunch. It's kind of like they're taking on the torch, the next step.

These pro-social, pro-environmental experiences constitute the means by which students develop a civic identity (Yates & Youniss, 1999 as cited in Chawla & Cushing, 2007). Venturing forth from school to engage in real-world, problem-based learning, children build a sense of individual and collective agency as they undertake meaningful endeavors and succeed (Chawla, 2009, p. 16). Through such experiences, kindergarten students in EUSD learned critical thinking skills, learned about the digestive tract, applied basic writing skills, and developed their presentation prowess, at the same time they discovered they were wholly capable agents, prepared to make important things happen in the world (Fullan, Quinn, & McEachen, 2018).

Conclusion

It turns out that developing citizenship for sustainability requires the very same practices that we know support individual well-being and engagement in learning. Living systems-minded school leaders cultivate the conditions in which students' well-being is high, and they are able to engage more deeply in meaningful learning. Along the way, they also contribute to improving their communities. To accomplish this fundamental shift in educational practice, living systems-minded leaders must prepare themselves to convince stakeholders that learning in nature does not result with students being lost in the weeds, so to speak. Rather, a growing body of evidence suggests that students thrive in nature—academically, cognitively, physically, and socially/emotionally. These beautiful experiences also lead students to "understand, use, and feel connected to the natural environment, [increasing the likelihood that they] will grow into adults who appreciate and protect its healthy, functional, and aesthetic properties" (Taylor, 2009, p. 327).

Leadership Design Challenges

1. **Protect recess time**.
   In accordance with the American Academy of Pediatrics, establish recess as a necessary break in the day for optimizing children's social, emotional, physical, and cognitive development. Protect recess as a child's personal time, never withholding it for academic or punitive reasons. Utilize The Crucial Role of Recess in School, *Pediatrics (2013)* as a faculty-wide discussion prompt. http://pediatrics.aappublications.org/content/131/1/183

2. **Plan, design, and plant a learning garden**.
   Meet with your school district facility and maintenance department to develop an approved plan for school learning gardens. Develop a garden agreement, signed by principals and the facilities department, which includes design specifications related to ADA compliance, siting requirements that take account of mechanical and plumbing systems, and maintenance guidelines for upkeep. Identify your initial Garden Champion (a parent, a local garden shop, a teacher) who has the capacity to teach teachers and students about gardening, irrigation, organic pest control, seed propagation, and construction. Break ground. For direction along the way, read *Learning Gardens and Sustainability Education: Bringing Life to Schools and Schools to Life* (Williams & Brown, 2012).

3. **Design and model an objective and standards-based unit that takes students out beyond the school walls**.
   Students might leave the school building to investigate their local community and collect data about the availability of fresh produce

in their neighborhood. They might visit corner groceries, drug stores, and even restaurants, asking challenging questions about food quality, availability, and cost. Or, they might investigate a natural ecosystem such as a local creek, pond, forest, meadow, or coastal area that is within a short walk from classrooms. Students become local experts, knowledgeable about their communities and better prepared to engage as citizens. Be sure to demonstrate how these outdoor learning experiences provide opportunities for literary and artistic inspiration, vocabulary learning, understanding complex math and science concepts, and language development.

 Learning From Living Systems-Minded Trailblazers

In a September 13, 2019 Ed Week article entitled *The Irrefutable Case for Taking Class Outside*, Kate Ehrenfeld Gardoqui*, former Maine Teacher of the Year, remarked that most teachers "assume outdoor learning is for fun, reflection, and personal growth, while rigorous academic work happens inside" (Gardoqui, 2019, p. 20). She further observed that this assumption persists even in the face of an expanding body of research "establishing that a connection to nature is essential for young people" (Gardoqui, 2019, p. 20). Both as a teacher, and in her current role as an instructional and leadership coach, Gardogui has dedicated herself to issues of equity, and in Gardoqui's mind, "When we address privilege and inequity, access to nature should be high on our list" (Gardoqui, 2019, p. 20).

As a high school teacher, Gardoqui taught students who had failed English or were at risk of dropping out. In her class, they read ecologist Tom Wessels' *Reading the Forested Landscape: A Natural History of New England* as their primary text, demonstrating their understanding of this college-level text through rigorous field exercises and writing assignments. Her students also read other leading nature writers and completed service-learning projects that required extensive research and subsequent communication of findings and outcomes. Gardogui explained, "By incorporating nature into our learning, we were not neglecting rigor in favor of fun. . . . I was fighting for equity by pushing my students to read sophisticated texts, write clearly and powerfully, and synthesize information" (Gardoqui, 2019, p. 20).

According to Gardogui, as we provide students opportunities to learn in nature, we support the goals of equity in terms of high standards, student voice, and responsive curriculum. As an instructional and leadership coach, Gardogui urges teachers to move learning and teaching "outside the box that is the classroom and into the natural world", at the same time she urges school leaders to "take the easy first step of supporting teachers in getting students outside" (Gardoqui, 2019, p. 20).

*Kate Ehrenfeld Gardoqui is a senior associate with the Great Schools Partnership and the cofounder of White Pine Programs, a nature-connection organization in southern Maine.

References

Bellafante, G. (2020). Schools beat earlier plagues with outdoor classes. We should, too. *New York Times*. Retrieved at www.nytimes.com/2020/07/17/nyregion/coronavirus-nyc-schools-reopening-outdoors.html.

Berto, R., Pasini, M., & Barbiero, G. (2015). How does psychological restoration work in children? An exploratory study. *Journal of Child and Adolescent Behaviour*, *3*(3), 200–209. doi:10.4172/2375-4494.1000200

Bølling, M., Otte, C. R., Elsborg, P., Nielsen, G., & Bentsen, P. (2018). The association between education outside the classroom and students' school motivation: Results from a one-school-year quasi-experiment. *International Journal of Educational Research*, *89*, 22–35. doi:10.1016/j.ijer.2018.03.004

Camassoa, M. J., & Jagannathan, R. (2018). Improving academic outcomes in poor urban schools through nature-based learning. *Cambridge Journal of Education*, *48*(2), 263–277.

Chawla, L. (2009). Growing up green: Becoming an agent of care for the natural world. *Journal of Developmental Processes*, *4*(1), 6–23.

Chawla, L. (2015). Benefits of nature contact for children. *Journal of Planning Literature, 30*(4), 433–452. doi:10.1177/0885412215595441

Chawla, L. (2018). The American Association for Teaching & Curriculum (AATC) keynote address: Nature-based learning for student achievement and ecological citizenship. *Curriculum and Teaching Dialogue, 20*(1 & 2), xxv–xxxix.

Chawla, L., & Cushing, D. F. (2007). Education for strategic environmental behavior. *Environmental Education Research, 13*, 437–452.

Emekauwa, E. (2004). *They remember what they touch: The impact of place-based learning in East Feliciana Parish*. Washington, DC: Rural School and Community Trust.

Fullan, M., Quinn, J., & McEachen, J. (2018). *Deep learning: Engage the world change the world*. Thousand Oaks, CA: Corwin Press.

Gardoqui, K. E. (2019). The irrefutable case for taking class outside. *Ed Week, 39*(5), 20.

Gelfand, L. (2010). *Sustainable school architecture*. Hoboken, NJ: John Wiley & Sons, Inc.

Gill, T. (2014). The benefits of children's engagement with nature: A systematic literature review. *Children Youth and Environments, 24*(2), 10–34.

Johnson, J. F., Uline, C. L., & Perez, L. (2019). *Teaching practices from America's best urban schools: A guide for school and classroom leaders* (2nd ed.). New York: Routledge/Taylor and Francis Group.

Jordan, C., & Chawla, L. (2019). A coordinated research agenda for nature-based learning. *Frontiers in Psychology, 10*, 1–10. doi:10.3389/fpsyg.2019.00766

Kelz, C., Evans, G. W., & Röderer, K. (2015). The restorative effects of redesigning the schoolyard: A multi-methodological, quasi-experimental study in rural Austrian middle schools. *Environmental Behavior, 47*, 119–139.

Korr, M. (2016). Fighting TB with fresh-air schools: RIMS' doctors launch a movement. *Rhode Island Medical Journal,* September Issue, 75–76.

Krapfel, P. (1999). Deepening children's participation through local ecological investigations. In G. A. Smith & D. R. Williams (Eds.), *Ecological education in action: On weaving education, culture, and the environment* (pp. 47–78). Albany, NY: State University of New York Press.

Kuo, F. E., & Faber Taylor, A. (2004). A potential natural treatment for attention-deficit/hyperactivity disorder: Evidence from a national study. *American Journal of Public Health, 94*(9), 1580–1586.

Kuo, M., Barnes, M., & Jordan, C. (2019). Do experiences with nature promote learning? Converging evidence of a cause-and-effect relationship. *Frontiers in Psychology,* 10(305), 1–9. doi:10.3389/fpsyg.2019.00305

Kuo, M., Browning, M., & Penner, M. L. (2018). Do lessons in nature boost subsequent classroom engagement? Refueling students in flight. *Frontiers in Psychology, 8,* Article 2253. doi:10.3389/fpsyg.2017.02253

Li, D., ·& Sullivan, W. (2016). Impact of views to school landscapes on recovery from stress and mental fatigue. *Landscape and Urban Planning, 148,* 149–158.

Lindholm, G. (1995). Schoolyards: The significance of place properties to outdoor activities in schools. *Environment and Behaviour, 27,* 259–293.

Lopez, R., Campbell, R., & Jennings, J. (2008). *Schoolyard improvements and standardized test scores: An ecological analysis.* Boston, MA. Retrieved from www.schoolyards.org/pdf/sy_improvements_test_scores.pdf

Louv, R. (2008). *Last child in the woods: Saving our children from nature deficit disorder.* Chapel Hill, NC: Algonquin Books of Chapel Hill.

Meier, D., Knoester, M., & D'Andrea, K. C. (Eds.). (2015). *Teaching in themes: An approach to schoolwide learning, creating community, & differentiating instruction.* New York: Teachers College Press.

Mustapa, N. D., Maliki, N. Z., & Hamzah, A. (2015). Repositioning children's developmental needs in space planning: A review of connection to nature. *Procedia: Social and Behavioral Sciences, 170,* 330–339. doi:10.1016/j.sbspro.2015.01.043

Noddings, N. (2013). *Education and democracy in the 21st century.* New York: Teachers College Press.

Selby, D. (2000). A darker shade of green: The importance of ecological thinking in global education and school reform. *Theory into Practice, 39,* 88–96.

Sobel, D. (2005). *Place-based education: Connecting classrooms and communities.* Barrington, MA: The Orion Society.

Sobel, D. (2008). *Childhood and nature: Design principles for educators.* Portland, ME: Stenhouse Publishers.

Taylor, A. (2009). *Linking architecture and education: Sustainable design of learning environments.* Albuquerque: University of New Mexico Press.

Toffler, S., & Louv, R. (2020). The urgent case for green schoolyards during and after COVID-19. *Green Schools Catalyst Quarterly*, September Issue, 68–75. Retrieved from https://catalyst.greenschoolsnationalnetwork.org/gscatalyst/september_2020/MobilePagedReplica.action?pm=2&folio=1#pg1

Ulrich, R. S. (1984). View through a window may influence recovery from cancer. *Science, 224,* 420–423.

Williams, D. R., & Brown, J. (2012). *Learning gardens and sustainability education: Bringing life to schools and schools to life.* New York: Routledge Taylor and Francis Group.

Williams, D. R., & Dixon, P. S. (2013). Impact of garden-based learning on academic outcomes in schools: Synthesis of research between 1990 and 2010. *Review of Educational Research, 83*(2), 211–235. doi:10.3102/0034654313475824

Yates, M., & Youniss, J. (Eds.). (1999). *Roots of civic identity.* Chicago: University of Chicago Press.

Zarghami, E., & Fatourechi, D. (2015). Impact of sustainable school design on primary school children's mental health and well-being. *International Journal of Advances in Agricultural and Environmental Engineering, 2,* 31–38.

STRATEGY

6

Play

Restore Nature Play Into the School Day

Opportunities for unstructured play have steadily decreased since the 1950s and unstructured play outdoors, in nature rather than on fabricated play equipment, has declined even more (Gray, 2011). This decline is associated with a plethora of negative effects limiting healthy development, overall well-being, and learning outcomes (Hanscom, 2016). Nature play, an emerging term in research literature but an ancient phenomenon, is child-led play in the natural world (Dankiw, Tsiros, Baldock, & Kumar, 2020). "Intrinsic to play are the following traits: curiosity; open-mindedness; high energy; eagerness to learn; and joy" (Cornell, 2017, p. loc 184), and time in nature ignites playfulness. The natural world includes more than the obvious wilderness areas that few can readily access. Pockets of nature, from small to vast spaces where plants and animals co-exist, can be found on school grounds, within cities, and across rural and suburban communities.

Greening schoolyards for the benefits of children and communities is a growing trend (Zaplatosch, 2019). Boston Public Schools (BPS) has intentionally redesigned their schoolyards to include more nature-centered play spaces, or playscapes. Lopez, Campbell, and Jennings' (2008) findings suggest academic benefits for BPS students were associated with the presence of improved outdoor spaces. and aligns well with an emerging wave of research documenting the value of nature play for children. Dankiw et al. (2020) found in their systematic review that

> ... nature play may have a positive impact on a range of children's health and developmental outcomes—specifically, PA [physical activity], health-related fitness, motor skill, cognitive learning, social and emotional development . . . [and] in terms of cognitive outcomes (play, learning, creativity), consistent positive improvements were reported. Such improvements pertained to: poetic writing, attention levels, punctuality, concentration in class, settling time, concentration after play, constructive play, associative play, imaginative play and functional play . . .
>
> (p. 12)

Rachel Carson understood and wrote about the benefits of nature play for children in the 1950s. The following quote was republished in a book, *The Sense of Wonder: A Celebration of Nature for Parents and Children*, after her death.

> A child's world is fresh and new and beautiful, full of wonder and excitement. It is our misfortune that for most of us that clear-eyed vision, that true instinct for what is beautiful and awe-inspiring, is dimmed and even lost before we reach adulthood. If I had influence with the good fairy who is supposed to preside over the christening of all children, I should ask that her gift to each child in the world be a sense of wonder so indestructible that it would last throughout life, as an unfailing antidote against the boredom and disenchantments of later years, the sterile preoccupations with things that are artificial, the alienation from the sources of our strength.
>
> (Carson, 1988, p. 44)

Carson spoke of wonder as a source of resilience that serves us throughout our lifetimes. Current research finds that nature play sparks wonder and wonder fuels learning and resilience (Schlembach, Kochanowski, Brown, & Carr, 2018). Cultivating student well-being and resilience, foundational to academic success, is a primary responsibility of educators. Incorporating nature play into the school day serves the whole child, their physical, social, emotional, and academic aims, as described next by Encinitas Union School District (EUSD) Superintendent Grey.

> You know, it's an interesting topic because I think in our roles, we get so hyper-focused on the learning and yes, there's play with learning, but often-times, when we are so focused on the learning, we forget about all the other pieces. So, when I think about Encinitas, one of our main focuses has been the whole child. I mean, consistently, that has been a value, a principle that we live by. And so, we're trying to think of all aspects of the whole child and including health and wellness, which has been and is one of our four pillars. With health and wellness, that's one of the big places where play comes in. . . . So, when I think about outdoor play particularly—it's making sure, first of all, our environments match what we believe. And so, we have been very intentional about making sure we have green spaces and making sure as much as possible there's some outdoor learning spaces that also can be used for play. We're in the stages right now of doing more natural outdoor play areas.

Restoring play into the school day takes time, intention, and con-tinuous evaluation. We suggest living systems-minded leaders (1) evaluate the school's play ecology; (2) define and schedule play into the school day; (3) observe children at play; (4) engage children in improving the school's play ecology; (5) advocate for nature play.

Evaluate the School's Play Ecology

Humanity's drive to play is innate. Children's play erupts spontaneously, if not stifled through an over-engineered system of rules, constraints, and negative consequences. A school's play ecology sits at the intersection of place, space, objects, rules, and supervision (Figure 6.1). Living systems-minded leaders ask questions such as: When and where do children have the opportunity to play during the school day? What are the conditions in which play occurs or could occur? As living systems-minded leaders consider their school's play ecology, they take inventory of the quality

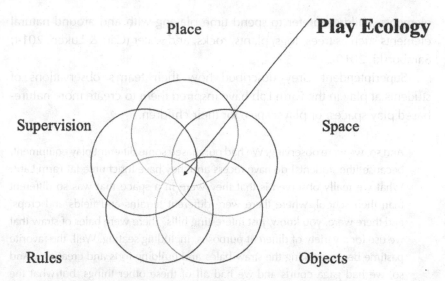

Figure 6.1 Your school's play ecology sits at the intersection of place, space, objects, rules, and supervision.

of the spaces in which children play and the objects on and with which children play. U.S. public schools sit on approximately 2 million acres of land (Filardo, 2016). Schoolyards often, but not always, include fairly large areas of undeveloped open space. This space may include native habitat such as forests, creeks, ponds, or other natural landscapes that easily facilitate nature play. Living systems-minded leaders see their whole campus as in service to learning and notice and pay special attention to these natural landscapes; these spaces will not simply exist as the untended "back forty." Of course, school administrators rarely have the time to do the work of developing and maintaining natural areas for student use and, thus, will depend upon partnerships with community members and organizations—a topic we take up in Strategy 8.

The reality is that most schoolyards are not ready made for nature play. More commonly, school yards include mowed lawns and playgrounds. At their most basic, playgrounds might be asphalt covered space with some painted lines for games like four square hopscotch, or gaga. Or, school playgrounds might include plastic and/or metal equipment set on a blank slate of dirt or rubberized substrate. This type of equipment provides a variety of opportunities for climbing and exploring physical limitations. Yet, when given the choice between human-made play elements and natural elements, children prefer to spend time playing with and around natural elements such as trees, logs, plants, rocks, and water (Carr & Luken, 2014; Samborski, 2010).

Superintendent Grey described how their team's observations of students at play in the Farm Lab have inspired them to create more nature-based play spaces, or playscapes, for their children.

> And so, we were observing. We had purchased some different play equipment, because [the students] do have recess and they have lunchtime [at Farm Lab]. What we really observed is that they were in a space that was so different than their school, where there were different terrains and fields and crops, and there were, you know, just interesting hills. There were bales of straw that we use for a variety of different purposes, including seating. Well, the favorite pastime became taking the straw bales and building forts and creations. And so, we had gaga courts and we had all of these other things, but what the children loved was either running through the fields and the crops or creating structures. It was that observation that made us realize, wow, we have an opportunity to create something pretty special and have students engaged in it. That can be a model for what we look at in our other settings. There is this kind of balance of nature and creativity and just being able to do something that isn't so predetermined as far as, you go up this slide and go down this slide. Here's your opportunity to see what you want to do with it [outdoor play space], and without the same protocols or rules to it.

Living systems-minded leaders expand opportunities for play to include nature play. For some, this may be as simple as welcoming a team of community members to join in a series of weekend cleanups to insure the otherwise ignored "back forty" is a safe space for play and exploration. For many others, it may mean developing a longer-term strategy for transforming asphalt into playscapes (Carr & Luken, 2014). Carr and Luken (2014) offered the following principles for guiding the design and development of playscapes:

- Playscapes elicit hands-on, multi-sensory, unique, and personal experiences for children where nature is the focus, not man-made materials.
- Areas within the playscape are designed to be open-ended with multiple and divergent uses.

- Materials and spaces are not designed to be used in pre-determined ways.

- Selected playscape plants and materials are ones that can be found in nature, preferably indigenous to the local landscape.

- Playscape materials provide opportunities to be touched, manipulated, dug, moved, picked, dammed, climbed, built, and experienced by children as they choose to do so.

- Playscapes are built to encourage risk-taking, investigation, language, sensory experiences, child-directed dramatic and themed play, and collaborative and active play.

(p. 74)

- We add to this list, playscapes must be accessible to all children. Wilson (2018) provided a detailed discussion of accessibility in her book, *Nature and Young Children: Encouraging Creative Play and Learning in Natural Environments*.

Safety spans the intersection of place, space, rules, and supervision. Of course, safety must be paramount as living systems-minded leaders integrate nature play into children's school day. Fencing the perimeter of playscapes with a single entrance/exit helps to keep children in and strangers out of the space (Carr & Luken, 2014). Rules ought to be simple, clear, and enforceable. Alter and Haydon (2017), in their review of effective classroom rules, suggested four rules as a reasonable target number because four items are readily remembered. Hanscom (2016) described a school whose outdoor play is governed by just two rules: "(1) Children have to be able to see an adult at all times, and (2) children need to show respect to the other children and adults" (p. 153). We suggest modifying rule two to include other living things as also worthy of everyone's respect. Regardless of the specific rules, taking time to discuss, interpret, and model what the rules mean for unique spaces and groups of children is critical.

Rules and associated supervision foster the play ecology's culture, the way we play here. As living systems-minded leaders consider the play ecology of their school, they engage conversation around questions like: What rules define when and how children play together? Is time to play a right or a privilege, and why? Are the same children deprived of time to play day in and day out? The very children seen as disruptive and challenging during academically focused periods of the school day may very well need the restorative effects of play, particularly nature play, the most (Kuo, Barnes, & Jordan, 2019). Rule-focused supervision and conflict resolution often frustrates children and exhausts teachers. In playscapes, teachers and students share an expanded sense of exploratory wonder for daily novelty and surprise, such as new insects visiting blooming flowers or water finding its way through a sand barrier (Kloos, Maltbie, Brown, & Carr, 2018). Fueled by wonder, teachers and students engage in less conflict and discover inspiring questions that ·motivate deeper classroom learning.

Define and Schedule Nature Play Into the School Day

Play, particularly nature play, serves academic aims. Children learn and build capacity for social, emotional, and academic gains through play (Dankiw et al., 2020). Playful time in nature restores attention and children's capacity to focus on and retain academic learning (Bagot, Allen, & Toukhsati, 2015) alongside so many other benefits discussed throughout this chapter. Scheduling nature play into the school day will support children's ability to focus and learn. Living systems-minded leaders understand the value of play and how to facilitate it. They engage their teaching staff in defining and scheduling play into their school day. If nature play is not already an essential element of a school's approach to learning, then focusing professional learning community efforts around nature play and clarifying the school's play philosophy is a powerful way to begin.

> We recommend the following books for further exploration, study, and discussion:
>
> - *Last Child in the Woods: Saving Our Children from Nature-Deficit Disorder* by Richard Louv (2005)
> - *Childhood and Nature: Design Principles for Educators* by David Sobel (2008)

- *Risk, Challenge, and Adventure in the Early Years: A Practical Guide to Exploring and Extending Learning Outdoors* by Kathryn Solly (2015)

- *Balanced and Barefoot: How Unrestricted Outdoor Play Makes for Strong, Confident, and Capable Children* by Angela J. Hanscom (2016)

- *Deep Nature Play: A Guide to Wholeness, Aliveness, Creativity, and Inspired Learning* by Joseph Bharat Cornell (2017)

- *Nature and Young Children: Encouraging Creative Play and Learning in Natural Environments* by Ruth Wilson (2018, 3rd Edition)

- *The Sky Above and the Mud Below: Lessons from Nature Preschools and Forest Kindergartens* by David Sobel (2020)

Each school's play ecology, philosophy, and approach will be unique. It will emerge from the collective process of exploring the concept of nature play, it's precursors, processes, and outcomes, alongside available places, spaces, and resources.

Superintendent Grey noted that EUSD schedules play into the school day in more traditional recess blocks. Emerging research confirms the value of these recess blocks (Bauml, Patton, & Rhea, 2020; Rhea & Rivchun, 2018). This time for play outside allows for

> a brain break (See Figure 6.2); it's scheduled for socialization; it's scheduled for outdoors and fresh air, but it's not necessarily scheduled as part of the instructional day to have kids purposefully play . . . and yes, we could strive to do more of that.

As Cornell (2017) makes clear, nature play is not just for children! Adults will benefit from this scheduled time spent in nature as well.

> As people age the natural openness, confidence, and adaptability of their early childhood years generally subside, to be replaced by such inhibitors as self-criticism and fear, inhibitors that often stifle an adult's ability to learn. Two of the

Figure 6.2 Students take a "brain break" to pet visiting animals during recess.

benefits of deep play—self-forgetfulness and living in the present—effectively quiet critical self-talk and other habits harmful to one's capacity to learn.

(loc. 240)

Observe Children at Play

Children spontaneously play in natural spaces. Left to their own exploration, they discover the joy of rolling down grassy hills, the accomplishment of balancing across a downed tree, the peace of just lying in the grass and watching clouds float by, and so much more. David Sobel, in his observational study of children at play described "seven play motifs" (Sobel, 2008):

1. Making forts and special places
2. Playing hunting and gathering games·
3. Shaping small worlds
4. Developing friendships with animals

5. Constructing adventures
6. Descending into fantasies
7. Following paths and figuring out shortcuts.

Other researchers have built on this work and created far more extensive lists of nature play behaviors. Kahn, Weiss, and Harrington (2018) published a list of 20 behaviors, what they call keystone interaction patterns, that capture the important ways in which children play in and interact with nature. Their list of behaviors extends Sobel's seven play motifs (2008) with more specificity. For each of their 20 behaviors, they detail evolutionary roots of the behavior, as well as present-day outcomes associated with human health, well-being, and flourishing. This line of research documents the ways in which children play and the benefits they reap from doing so. Expanding educator's capacity to see and understand nature play behaviors will lead to greater appreciation for these behaviors and the developmental value they hold.

Living systems-minded leaders recognize the importance of providing children opportunities to play in nature (See Figure 6.3). These leaders maintain a high level of curiosity about *how* the children in their care play and the benefits that follow. Hanscom (2016) infused her book, *Balanced and Barefoot*, with observations of today's children who are generally more deprived of nature play than previous generations. She noted increased referrals of children for occupational therapy, physical therapy, and speech therapy across many research studies. This research aligned with her own retrospective interviews of veteran teachers: "Over the years, they had noticed a slow decline in gross and fine motor ability, safety awareness, self-control, attention, and coordination" (Hanscom, 2016, p. 10). The solution? Provide children with more opportunities for nature play; it is absolutely critical for their development, well-being, and learning. Children not only need nature play, they prefer nature play (Samborski, 2010).

Observing children at play in EUSD led their team of educators to notice that children are drawn to play with natural objects, in natural spaces, as described in the previous section. As a result, EUSD's facilities master plan, completed prior to the COVID-19 pandemic, prioritizes improved outdoor learning spaces across all of their campuses. Their intent, according to Superintendent Grey, is to create more "outdoor learning spaces that look and feel like nature," that go beyond their extensive array of school gardens and more intentionally create opportunities for play and academic learning outside.

Figure 6.3 Children explore among tall grasses.

Engage Children in Improving the School's Play Ecology

When children have opportunities to engage with nature, they fall in love. This powerful finding emerged from listening to 68 first- through fifth-grade students describe their own experiences and relationships with nature (Kalvaitis & Monhardt, 2015). Their study revealed the deep understanding children hold about their relationship with nature.

> The results indicate that children have a positive deep-seated appreciation for nature and this fondness for nature is directly tied to their lived experiences in nature. Children simply love nature. The predominant themes from the study clearly indicate that nature provides children with opportunities for play/

work, home, beauty, freedom, learning, and relaxation as well as a critical life support system. Interestingly, the children did not take these things for granted and were aware of the real benefits nature provides them.

(p. 14)

Children also know what they find fun, exciting, and engaging. If asked, they will tell you (Almers, Askerlund, Samuelsson, & Waite, 2020). Living systems-minded leaders invite students to be partners in design, maintenance, and improvement of their play ecology. Students in EUSD are key stakeholders and designers in their efforts to expand and improve nature play spaces. As Superintendent Grey explained,

> What we've done with Farm Lab, all along the way, is involve students in the process and in the design and then the actual implementation of their design. So, our next one that our Farm Lab director has been working on is creating an outdoor play area that's completely student designed, and student led. It started with getting students together, talking about the benefits, what did they see, and giving them some examples. And then, then the idea is for them to create, plan the space, and resource out any of the materials that are needed. We did the same thing for our maker space and our kitchen. . . . The outdoor playground is the next one and it is supposed to be all natural with boulders, logs, whatever it might be that would be engaging for students.

When children have voice, when they have the opportunity to engage with adults in making decisions about aspects of school that affect them directly, they learn life skills; they develop self-esteem; they practice democratic skills; they develop healthy student-adult relationships; and they contribute to more positive school cultures (Mager & Nowak, 2012).

Advocate for Nature Play

Many of the neighborhoods in EUSD, like so many school districts across the United States, are so densely populated that access to high-quality green space is scarce, especially for children living in higher-poverty neighborhoods (Rigolon, 2016). This inequity in access to green space may very well compound observed gaps in learning outcomes across groups of children already underserved in so many ways. Researchers have recently engaged in studying the greenness-academic achievement link (G-AA). Kuo, Browning, Sachdeva,

Lee, & Westphal (2018) explored this G-AA link in an urban district (Chicago, IL) that serves predominantly children of color living in poverty; 90% of students qualify for free lunch and white students account for only 10% of the student population. Their study found that the presence of trees on school property were significantly associated with better student math and reading performance and grass and shrub cover were not. These findings align with research cited throughout this book that emphasizes the beneficial effects of time spent in nature. Unfortunately, they also found fewer trees existed in neighborhoods where most students of color and students living in poverty resided, thus highlighting a critical environmental justice issue.

EUSD is working to green their schoolyards to better serve their neighborhoods, as described next by Superintendent Grey.

> It is important aesthetically for our schools so that they feel like a part of the community, that they don't feel like an institution, that there are places where students feel comfortable. When we look at our community as a whole, we do have many areas that are more of your traditional neighborhood feel, not a neighborhood with a lot of trees, but neighborhoods in California that are high-density housing. And so, our schools many times provide the entire play-ground or the outdoor space for the whole neighborhood. We have several of our schools that are complete open campuses because we want the neighbors to feel like this is a space for them. It's a community place where they can gather. So, making sure our schools have that park feel is important because, in some cases, it is the only park in the area or the only green space in the area. (Figure 6.4 pictures an open campus in EUSD.)

Children in Encinitas neighborhoods increasingly have access to high-quality green space, because their school district has made it a priority. Their advocacy for nature play takes the form of providing equitable access to engaging natural spaces and continuously improving the quality of these spaces for their students and community. Living systems-minded leaders whose schoolyards remain void of engaging and restorative natural features do well to advocate for investment in their outdoor spaces so that their children have healthy places to play. In their (2018) study of renovated schoolyards, Bates, Bohnert, and Gerstein concluded,

> The current study builds on existing literature that has shown benefits of green schoolyard renovations to PA [physical activity], prosocial behavior, and safety, and provides additional evidence that renovated green schoolyards in

Figure 6.4 Natural places on school campuses provide high quality space for community.

low-income urban areas serve as a beneficial context of development for at risk youth. . . . Taking these results in the context of other literature leads us to conclude that investing in built environments, particularly green schoolyards, may be an effective and enduring way to promote positive development outcomes among school-age youth, especially those living in low-income urban neighborhoods with limited other resources.

(p. 8)

Conclusion

Living systems-minded leaders value nature play for the array of benefits that it provides students. Children learn through play. Their time in nature contributes to their well-being and restores their capacity to focus and attend to academic learning in the classroom. Each school's play ecology is a unique integration of place, space, objects, rules, and supervision. Living systems-minded leaders are intentional about each aspect of their play ecology because they understand nature play serves academic aims and is worthy of dedicated time in the school day. They also understand the importance of developing high-quality playscapes that include natural features with which to interact—everything from boulders for climbing to sticks and leaves for creative play. They partner with local groups to fund, design, and build more nature-inspired play spaces, while also including students in imagining and planning. These playscapes can be a resource for the community, even serving as a form of advocacy for nature play beyond school hours. As Superintendent Grey notes, "What we have found anecdotally, and I know research supports it as well, is that students thrive when they go outside."

Leadership Design Challenges

1. **Develop a play leadership committee that includes parents and students.** Ask for volunteers who are interested in learning more about the research-based benefits of nature play and designing a plan for increasing nature play at your school.

    a. Begin with a book study. We recommend Angela J. Hanscom's book, *Balanced and Barefoot: How Unrestricted Outdoor Play Makes for Strong, Confident, and Capable Children*, for its grounding in interdisciplinary sciences and highly practical and accessible approach.

    b. Following the book study, evaluate your school's play ecology using the framework presented in this chapter—place, space, objects, rules, supervision. Identify areas for improving your school's play ecology in the short and long term.

    c. As you consider your play ecology, how would you characterize the physical space for play? Along a continuum from 100% asphalt to 100% natural habitat, what does your school campus offer students? What could it offer students? Develop an action plan for designing, building, and maintaining more natural playscapes.

    d. Develop at least one team of teachers and students who are willing to conduct action research projects for studying the effects of play. Co-create these studies and implement them. Learn from students about play and notice how it relates to their engagement in learning (Green, 2015; Mitra & McCormick, 2017; Ozer, 2017; Ozer et al., 2020).

e. Continuously learn and improve the quality of nature play opportunities at your school.

Learning from Living Systems-Minded Trailblazers

Forest Schools, typically nature immersion preschools, are spreading rapidly across the United States (NAAEE, 2017). Less common are nature immersion opportunities for children in elementary schools. The Brooklyn New School, a public school not zoned for a particular neighborhood in New York City, admits students via annual lottery. They have been participating in New York City's Diversity in Admissions plan for at least five years, prioritizing admission to students living in poverty. They serve 685 students PK–5th grade, 33% of whom qualify for free or reduced lunch. Their student body includes 23% who identify as Hispanic, 18% who identify as Black, 13% who identify as multiracial, 5% who identify as Asian, and 41% who identify as white.

Fulbrecht documents Brooklyn New School activities on his blog, *BNS Forest and Shore*. Their pre-K, kindergarten, first-grade, and second-grade students have the opportunity to visit nearby Prospect Park once each week. Fulbrecht described these weekly field trips,

> they are given the opportunity to freely explore and engage with nature. They quickly notice changes in the environment with each trip. They note the colors of the leaves, the textures of tree bark, the sounds of the wind and rain, the presence of birds, squirrels, insects and worms, the consistency of mud and sand, the shimmer of ice and snow, the solidity of rock, the resilience of wood. The forest environment offers endless variety and ignites children's imaginations. Children engage in both cooperative play and individual inquiry, exploring together, testing their abilities, following their curiosity.
>
> The lessons learned in the forest arrive through the body. They are learned through *doing* rather that *listening*. The experience of dragging a

heavy object through mud in falling snow or holding a wriggling worm in your cupped hands becomes the concrete basis for understanding the physics of inertia or the cycles of life. These experiences are also the basis for understanding one's place within nature and for recognizing the interrelatedness of all things.

(Fulbrecht, n.d.)

The pictures and videos throughout Fulbrecht's blog show children in all kinds of weather, dressed appropriately, and fully engaged in exciting exploration and play. His December 17, 2018 post, 'Ritual, Memory, and Mud,' described their tradition of starting each visit to the park by circling a large tree, observing it, and sharing their observations. Children see changes from visit to visit, season to season. Fulbrecht explained further,

This simple ritual of forming a circle and noticing the changes in one individual tree will, hopefully, become one of the strongest memories for the children as they themselves grow and change. It is a ritual of quiet observation, a peaceful communion with the wonder that is life. If we get to know this one tree as a friend it may change the way we see all trees, just as getting to know one person as a friend can change the way we see all people.

The Brooklyn New School and their Living Systems-Minded Trailblazers are cultivating deep connections among their students, teachers, and nature through play, inquiry, and exploration.

References

Almers, E., Askerlund, P., Samuelsson, T., & Waite, S. (2020). Children's preferences for schoolyard features and understanding of ecosystem service innovations: A study in five Swedish preschools. *Journal of Adventure Education and Outdoor Learning*, 1–17. doi:10.1080/147 29679.2020.1773879

Alter, P., & Haydon, T. (2017). Characteristics of effective classroom rules: A review of the literature. *Teacher Education and Special Education*, *40*(2), 114–127. doi:10.1177/0888406417700962

Bagot, K. L., Allen, F. C. L., & Toukhsati, S. (2015). Perceived restorativeness of children's school playground environments: Nature, playground features and play period experiences. *Journal of Environmental Psychology, 41*, 1–9. doi:10.1016/j.jenvp.2014.11.005

Bates, C. R., Bohnert, A. M., & Gerstein, D. E. (2018). Green schoolyards in low-income urban neighborhoods: Natural spaces for positive youth development outcomes. *Frontiers in Psychology, 9*, 805. doi:10.3389/fpsyg.2018.00805

Bauml, M., Patton, M. M., & Rhea, D. (2020). A qualitative study of teachers' perceptions of increased recess time on teaching, learning, and behavior. *Journal of Research in Childhood Education*, 1–15. doi:10.1080/02568543.2020.1718808

Carr, V., & Luken, E. (2014). Playscapes: A pedagogical paradigm for play and learning. *International Journal of Play, 3*(1), 69–83. doi:10.1080/21594937.2013.871965

Carson, R. (1988). *Sense of wonder*. New York: Open Road Integrated Media.

Cornell, J. B. (2017). *Deep nature play: A guide to wholeness, aliveness, creativity, and inspired learning*. Nevada City, CA: Crystal Clarity Publishers.

Dankiw, K. A., Tsiros, M. D., Baldock, K. L., & Kumar, S. (2020). The impacts of unstructured nature play on health in early childhood development: A systematic review. *PLoS One, 15*(2), doi:10.1371/journal.pone.0229006

Filardo, M. (2016). *State of our schools: America's K–12 facilities 2016*. Washington, DC. Retrieved from www.centerforgreenschools.org/state-our-schools

Fulbrecht, B. (n.d.). Why forest and shore school? *BNS Forest and Shore*. Retrieved from https://bnsforestandshore.blogspot.com/p/why-forest-and-shore-school.html

Gray, P. (2011). The decline of play and the rise of psychopathology in children and adolescents. *American Journal of Play, 3*(4), 443–463.

Green, C. J. (2015). Toward young children as active researchers: A critical review of the methodologies and methods in early childhood environmental education. *The Journal of Environmental Education, 46*(4), 207–229. doi:10.1080/00958964.2015.1050345

Hanscom, A. J. (2016). *Balanced and barefoot: How unrestricted outdoor play makes for strong, confident, and capable children*. Oakland, CA: New Harbinger Publications, Inc.

Kahn, P. H., Jr., Weiss, T., & Harrington, K. (2018). Modeling child-nature interaction in a nature preschool: A proof of concept. *Frontiers in Psychology, 9*, 835. doi:10.3389/fpsyg.2018.00835

Kalvaitis, D., & Monhardt, R. (2015). Children voice biophilia: The phenomenology of being in love with nature. *Journal of Sustainability Education, 9*, 1–21.

Kloos, H., Maltbie, C., Brown, R., & Carr, V. (2018). Listening in: Spontaneous teacher talk on playscapes. *Creative Education, 9*(3), 426–441. doi:10. 4236/ce.2018.93030

Kuo, M., Barnes, M., & Jordan, C. (2019). Do experiences with nature promote learning? Converging evidence of a cause-and-effect relationship. *Frontiers in Psychology, 10*. doi:10.3389/fpsyg.2019.00305

Kuo, M., Browning, M., Sachdeva, S., Lee, K., & Westphal, L. (2018). Might school performance grow on trees? Examining the link between "greenness" and academic achievement in urban, high-poverty schools. *Front Psychol, 9*, 1669. doi:10.3389/fpsyg.2018.01669

Lopez, R., Campbell, R., & Jennings, J. (2008). *Schoolyard improvements and standardized test scores: An ecological analysis.* Boston, MA. Retrieved from www.schoolyards.org/pdf/sy_improvements_test_scores. pdf

Mager, U., & Nowak, P. (2012). Effects of student participation in decision making at school: A systematic review and synthesis of empirical research. *Educational Research Review, 7*(1), 38–61. doi:10.1016/j. edurev.2011.11.001

Mitra, D., & McCormick, P. (2017). Ethical dilemmas of youth participatory action research in a democratic setting. *International Journal of Inclusive Education, 21*(3), 248–258. doi:10.1080/13603116.2016.1260835

North American Association for Environmental Education (NAAEE). (2017). *Nature preschools and forest kindergartens: 2017 national survey.* Washington, DC: NAAEE. Retrieved from https://naturalstart.org/sites/ default/files/staff/nature_preschools_national_survey_2017.pdf

Ozer, E. J. (2017). Youth-led participatory action research: Overview and potential for enhancing adolescent development. *Child Development Perspectives, 11*(3), 173–177. doi:10.1111/cdep.12228

Ozer, E. J., Abraczinskas, M., Voight, A., Kirshner, B., Cohen, A. K., Zion, S., . . . Freiburger, K. (2020). Use of research evidence generated by youth: Conceptualization and applications in diverse U.S. K–12 educational settings. *American Journal of Community Psychology.* doi:10.1002/ajcp.12425

Rhea, D. J., & Rivchun, A. P. (2018). The LiiNK Project®: Effects of multiple recesses and character curriculum on classroom behaviors and listening skills in grades K-2 children. *Frontiers in Education, 3.* doi:10.3389/feduc.2018.00009

Rigolon, A. (2016). A complex landscape of inequity in access to urban parks: A literature review. *Landscape and Urban Planning, 153,* 160–169. doi:10.1016/j.landurbplan.2016.05.017

Samborski, S. (2010). Biodiverse or barren school grounds: Their effects on children. *Children, Youth and Environments, 20*(2), 67–115.

Schlembach, S., Kochanowski, L., Brown, R. D., & Carr, V. (2018). Early childhood educators' perceptions of play and inquiry on a nature playscape. *Children, Youth and Environments, 28*(2), 82–101.

Sobel, D. (2008). *Childhood and nature: Design principles for educators.* Portland, ME: Stenhouse Publishers.

Sobel, D. (2020). *The sky above and the mud below: Lessons from nature preschools and forest kindergartens.* St. Paul, MN: Red Leaf Press.

Wilson, R. (2018). *Nature and young children: Encouraging creative play and learning in natural environments.* New York: Routledge/Taylor and Francis Group.

Zaplatosch, J. (2019). Green schoolyards: Benefits, models, and tips for successful outcomes. *Green Schools Catalyst Quarterly, 6*(1), 50–59. Retrieved from https://catalyst.greenschoolsnationalnetwork.org/gscatalyst/march_2019/MobilePagedReplica.action?pm=2&folio=50#pg50

Care

In their own particular place on Earth, living-systems-oriented school leaders hold themselves to account as their schools' lead learners. They take time to clarify their own sense of purpose as educators and challenge themselves to investigate the implications of current societal and environmental challenges for their work as 21st-century school leaders. This reflection prompts them to consider their role as civic leaders across the social, economic, *and* ecological systems within which their schools are nested. The potential scope of their responsibilities can seem overwhelming, and so, they engage others in crafting a laser-focused vision for their work, a vision grounded in intimate knowledge of, and *CARE* for, the place their students call home. In this way, they are able to Cultivate and model Awareness, Responsibility, and Empathy. When people are able to connect their daily work to meaningful, purposeful aims, motivation soars. They feel passionate about their contribution to their vision for a healthy, flourishing community.

Model

Cultivate and Model Care, Awareness, Responsibility, and Empathy

One primary lesson we learn from nature is that of interdependence. Humans are dependent on natural systems for fresh air, clean water, nutritious food; our life systems benefit from nature's healthy functioning. And, nature's healthy functioning depends on how humans approach living on this planet. The sustainability movement aims to address this interdependence across socio-ecological communities by realigning human activities with the laws of nature. One's commitment to living more sustainably, more in harmony with nature's laws, is a mindset (or set of mental models) that expands on this lesson of interdependence and expresses a more expansive ethic of care. Dr. Baird, former superintendent of Encinitas Union School District (EUSD), shared this about sustainability:

> Sustainability. That's a loaded word. I think sustainability, in many ways gets mislabeled as resources, and dollars, and things like that. I think it's really about mindset. Sustainability is about how we look at things. . . . Sustainability mindset is about making sure, in terms of developing learners, that we're developing learners that care about their world, care about each other, and do everything in their power to support that. Sustainability may not be this particular program or this particular grant or whatever, but it's that overall mindset.

Educational leaders deepen their mindset for sustainability as they expand their capacity to CARE about their local communities, to CARE about their world, and to CARE about each other, through cultivating awareness, responsibility, and empathy. This strategy outlines four steps that living systems-minded school leaders take to serve the best interests of students during this period of rapid social, economic, and environmental change. They cultivate and model an (1) Expansive Ethic of Care, (2) Awareness, (3) Responsibility, and (4) Empathy. These steps are mutually reinforcing, as illustrated in Figure 7.1.

Cultivate and Model an Expansive Ethic of Care

In *Leadership for Green Schools* (Kensler & Uline, 2017), we argued that the purpose of schooling is to educate every student to their full potential in ways that prepare and empower them to respond to the needs of the 21st century and beyond. To the degree schools provide opportunities for students to learn deeply and think critically, they prepare students to respond intelligently to these needs. Such schooling fuels students' efforts to make a positive difference both locally and globally, presently and throughout their lives.

A school district's efforts to meet students' academic, social-emotional, and health-related needs cannot be separated from their efforts to meet the needs of their communities and the planet, today and tomorrow. Living

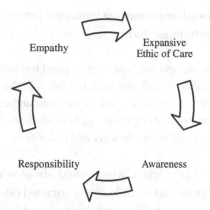

Figure 7.1 The elements of the CARE model are mutually reinforcing.

systems-minded leaders accomplish these interdependent and mutually reinforcing aims through implementation of leadership and management strategies that model an *expansive* ethic of care. We emphasize 'expansive' because so much of the literature related to ethic of care, especially in educational leadership, focuses exclusively on caring for students presently in schools (Bass, 2020; Louis, Murphy, & Smylie, 2016). Starratt (2014) described ethic of care as,

> A school community committed to an ethic of caring will be grounded in the belief that the school as an organization should hold the good of human beings within it as sacred. This ethic reaches beyond concerns with efficiency, which can easily lead to using human beings as merely the means to some larger purpose of productivity, such as an increase in the district's average scores on a standardized test or the lowering of per-pupil costs.
>
> (p. 56)

Living systems-minded leaders extend their circle of care beyond the students within their walls to include all of today's children, tomorrow's children, and our planet. Their behaviors and actions lead students to feel valued and cared for and teach students all the ways they can care for themselves, each other, and the world beyond their school campuses.

Educational leaders in Encinitas Unified School District, EUSD, embrace this expansive ethic of care and speak with clarity about the importance of striving for improved academic performance through authentic, meaningful learning opportunities. Former superintendent, Dr. Baird explained,

> I think good learning and teaching lead to good test scores. . . . We do look at data, we look at lots of different data, but that isn't the driver for us. My job is to provide opportunities for learning in the most authentic, meaningful way we can. Opportunities that engage not just our students, but all the learners—that's teachers, that's all of us—in ways that make sense.

One of the EUSD principals, when asked about what effective school principals need to know and be able to do, focused on a leader's capacity to develop active, authentic, sustainability-focused learning cultures that engage all members in caring about, and acting on behalf of, their local communities.

> I really think that they have to have a mission, or a purpose, and that sustainability needs to be a part of that. I just really believe that the purpose of schools is preparing students, and even the adults that work in the schools, to be able to learn together and to use that learning to make a positive impact on their communities.

Bottery (2014) framed this larger purpose of schooling in terms of current generations, future generations, and our planet, asking the question "If we have not inherited the world from our ancestors but, rather, have borrowed it from all life to come, in what condition will we return it?" (p. 91). Bottery posed this distinction as a primary ethical concern for 21st-century educational leaders. Within EUSD, living systems-minded leaders and teachers heighten their students' attention to the levels of social, economic, environmental, and individual well-being present in their community. Authentic learning experiences have students taking responsibility for that which needs improvement and cultivating deep empathy for themselves, each other, and the planet. Former Superintendent Baird described how students and teachers across the district were challenged to identify a problem, assume responsibility by engaging in problem solving, and cultivate expanding circles of empathy.

> We're pretty purposeful about allowing students and staff to follow their passion and turn it into meaningful work. For instance, right now, I have a challenge out to all of our students and teachers to reduce the plastic packaging

associated with lunchtime. We have been getting rid of plastic packaging, but then students themselves raised the issue of plastic sporks. Classes all over the district are having this conversation through a format called the Learning Quest. They are coming up with ideas about how to get food from the plate to the mouth without using a plastic spork. It doesn't matter whether you're a kindergartener or whether you're someone else in our organization. If you have a good idea, we try to shine some light on it and then take action.

Living systems-minded leaders in EUSD, children to adults, are encouraged to practice this more expansive ethic of care; to actively notice what they notice across personal, social, and environmental domains; and then to act on their awareness for continuous improvement.

Cultivate and Model Awareness

The 21st century is a time of rapid change and a time that demands rapid change, if we are to successfully address the many needs across nested spheres encompassed by the sustainability movement; needs across interdependent spheres of environmental, social, economic, and individual well-being concerns (AtKisson, 1999). These needs call for deep changes, not simple changes, and they touch all aspects of our lives (Klimek & AtKisson, 2016). Without awareness of issues and opportunities for learning and change, little will actually change. Research has repeatedly shown the importance of childhood experiences for cultivating lifelong awareness of, and engagement with, socio-ecological issues (Chawla, 1998, 2009; Derr, 2020). Thus, a focus on cultivating awareness among students and adults is key to managing current resources responsibly and also educating for a sustainable future.

There are so many examples of students, teachers, administrators, and community members raising awareness of current community and worldwide needs at EUSD. Everywhere one looks, there is evidence of teacher- and student-led initiatives for positive change, both locally and globally. While interviewing one of the principals, two fifth-grade boys politely interrupted to update their principal on their progress towards expanding their book club. The principal noted that she and teachers encourage students to notice what needs changing and to then generate ideas for leading change efforts. Often students' efforts to lead change first require raising their school community's awareness of the need for change. A fourth-grade teacher described one example related to improving recycling behaviors at their school.

> So, the fourth-graders at our school are in charge of the recycling program. [Each group of students is responsible for a different area of the school.] Every week, there are two days of collections, and they go through and collect data on contamination in the trash and/or the recycling. Then they come together as the whole grade level and they have to come up with a plan to make [the people in] their areas more aware or to do a better job recycling through reduced contamination.

Not only are students at EUSD empowered to act on their increased awareness of local, regional, and global needs, but they are also being taught strategies for leading change in their own communities or spheres of influence, including the importance of cultivating awareness prior to expecting people to take responsibility.

Cultivate and Model Responsibility

When students learn to act in service to their awareness of needs, they naturally embrace greater responsibility for their actions and engage further in learning and leading change (Monroe, Plate, Oxarart, Bowers, & Chaves, 2019). A principal shared this story about an elementary student who embraced a sense responsibility for addressing bullying school-wide.

> This girl, she's actually a sixth-grader now, but she was noticing that one of the boys in her class was being bullied. Kids were making fun of him and picking on him a lot, so on her own she told her teacher, "I want to talk to the kids about why what they're doing is called bullying, and why we shouldn't be doing it and the effect that it has on people, on the bully and on the student who's being bullied." So, just on her own, she created a PowerPoint presentation for her class and invited me to watch her presentation. From there, her teacher, at a staff meeting, told all the other teachers about it, and teachers signed up to have her come present.

The administration and teachers empowered this student to act in service to a need she noticed in her community. They made it possible for her to take responsibility and, in so doing, they cultivated her capacity for continued leadership in her community.

One of the principals called attention to a bulletin board highlighting students' stories of leading change beyond the school community. She explained,

> One of the things we've started doing, too, is documenting when students take action outside of school, because the cool thing about encouraging them to [lead necessary changes] in school is that now we have so many kids [for whom] it's just who they are. They do it on their own.

And students continue to lead after they leave EUSD and enter secondary schools. Dr. Baird, former superintendent of EUSD, shared stories of change led by former EUSD students. Upon entering secondary schools, they discovered there were no green teams. They felt the absence of opportunities for students to actively engage in leading change. So, with confidence and courage developed in elementary school, these EUSD graduates created green teams and took responsibility for continuing their work of making their communities and the world better.

Students at EUSD are developing action competence, that is, "the capacity to be able to act, now and in the future, and to be responsible for one's actions" (Jensen & Schnack, 1997, p. 175). Jensen and Schnack (1997) further explained,

> Therefore, there is a need for a form of teaching from which pupils acquire the courage, commitment and desire to get involved in the social interests concerning these subjects (naturally based on understanding and insight). They have to learn to be active citizens in a democratic society.
>
> (p. 164)

And they continued,

> But it can also be said that democracy is participation. In a democracy, the members are not spectators, but participants; not equally active participants in everything all the time, naturally, but always potential participants who decide for themselves in what and when they will be involved. Education for democracy is thus also socialization and qualification for the role of being a participant.
>
> (p. 165)

161

A more recent study, investigating the circumstances from which action competence for sustainability emerged, found the following conditions to be important (Almers, 2013):

- Emotional engagement in relevant and important issues
- Core values that encouraged individual and collective action on environmental and social issues
- Adults acting as role models, leading and acting consistent with stated values
- Opportunities to act with positive support and reinforcement from adults
- Trust and faith from and in adults
- A sense of belongingness within community

EUSD provides students with conditions consistent with Almers' (2013) findings. As they engage students in the real work of making their local and global communities better, EUSD educators model what it means to live in, and contribute to, a healthy democracy, while also providing students practical opportunities to lead the real work of change. They are learning to advocate for themselves and others. They are learning to speak up and act responsibly. One of the EUSD principals emphasized,

> That's one of the biggest things I tell my kids. "If I could teach you anything before you leave this campus, it would be how to advocate for yourself and how to speak up, because that's a life skill that we need."

Almers (2013) emphasized the important role that trusting students, to take responsibility for themselves and others, plays in cultivating action competence for sustainability related initiatives.

> The trust and faith, shown by emotionally important adults, emphasized the desire to respond by assuming responsibility. . . . Trust and faith from others may [also] strengthen a perception of self-trust which, in turn, can promote the will and courage to act, even while fighting an uphill battle.
>
> (p. 123)

Cultivate and Model Empathy

Empathy, the capacity to feel with others what they are feeling, is a natural part of being human; it appears as early as toddlerhood. Social emotional learning efforts identify empathy as a critical capacity to cultivate in children, alongside self-awareness, self-management, social skill, and decision-making (Eklund, Kilpatrick, Kilgus, Haider, & Eckert, 2019; Goleman & Senge, 2014). Further, we know powerful empathic triggers affect children on a daily basis. Children today are metaphorically swimming in information about social injustice and unrest, biodiversity loss, and climate change, all alongside awareness of their own and/or their peers' immediate needs associated with bias, poverty, and hunger. Mental health professionals report high rates of anxiety and depression among young people (Ghandour et al., 2019), and some credit climate anxiety or eco-anxiety as emerging contributors (Clayton, 2020; Taylor & Murray, 2020). Importantly, researchers are also learning that empathy without agency can fuel anxiety, depression, and withdrawal (Brown et al., 2019).

As educators cultivate empathy, if they are not also cultivating action competence, they may unwittingly be fueling increased anxiety and depression amongst their student body. Living systems-minded leaders equip students at EUSD to face local and global challenges—individual, social, and environmental—with agency and efficacy. Emerging research suggests this sense of agency and efficacy may very well serve as protective armor against anxiety due to empathic distress (Brown et al., 2019). As students at EUSD experience opportunities to take responsibility and act on their sense of empathy for other living things, human and nonhuman alike, they undoubtedly build their empathic capacity to notice and respond to widening circles of need; their ethic of care becomes more expansive. Three stories, told by two EUSD principals, illustrate the power of empathy where action competence is also nurtured.

In the first story, a student who loved drawing cartoons observed a need that sparked an idea, not only serving himself well, but also meeting a need for others. As told by his principal,

> This student is super into drawing cartoons. So, he said to me, "I really think that there are kids who don't like playing on the playground during recess time," and he asked, "Could we instead draw cartoons?" I responded, "Sure, if you want to start a cartooning club." On his own, he took care of securing the resources for drawing paper and art supplies. He found a teacher that would be willing to let them use her room during lunch. Then he advertised and made announcements. Now he and others have a cartooning club.

This student felt less excited about recess; he was able to acknowledge his discomfort and act on it, demonstrating self-empathy. He trusted his principal enough to be vulnerable and offer a possible solution. She trusted his capacity to act in service to a need he noticed in himself and saw that others might share. Although the principal did not share specifics, readers can imagine their own students who hang on the sidelines during recess for any long list of possible reasons. How often do students like these have empathy for themselves, advocate for themselves, and act with confidence to make things better for themselves and others? Educational leaders at EUSD are cultivating the conditions that facilitate this virtuous cycle of learning and healthy culture development.

In another example, an EUSD principal described a student who responded directly to her peer's need for comfort.

> I was in the nurse's office one day where an injured student was really upset. There was another girl who happened to be sick and was also in the nurse's office. I observed this student saying to the injured student, "It's okay, just use your yoga breath" to help the child calm down. They started breathing together just on their own. I was like, "What in the world!?" You would never see that anywhere else.

Students at EUSD take yoga classes regularly. They learn the power of their own breath to calm their thoughts and develop intentional mindfulness practices. This student's compassionate response to her peer aligns with emerging research that suggests a strong positive correlation between teaching students mindfulness practices and increased expressions of empathy and compassion (Cheang, Gillions, & Sparkes, 2019). ·

Although discussions of empathy often focus on interpersonal relationships, empathy is showing up as critical to human-nature relationships, as well (Brown et al., 2019). As we have discussed throughout this book, the sustainability movement is not only about interdependence among human communities, both locally and globally, but it is also about humanity's relationship with nature. The sustainability movement seeks solutions to severe social, as well as environmental, challenges. Living systems-minded leaders cultivate and model empathy for all of nature, alongside their human-centered efforts. They are developing aware, responsible, and action-oriented citizens.

A valuable example of sparking intense empathy, which could have induced anxiety and despair, occurred in a kindergarten class in EUSD.

You may recall from our discussion of Strategy 5 LEARN, that the class visited a local environmental center and learned about the destructive impact that plastic waste from their community was having on wildlife. Graphic images of dead birds with plastic waste in their stomachs triggered extreme empathy for the wildlife and deep concern that they take action to reduce plastic waste in their school. These kindergarteners insisted that they do something to help. Their principal shared that, although advocacy was not originally part of the lesson plans, she and the teachers honored their students' need to act and facilitated their leading a district-wide effort to rid the district of plastic straws.

> The thing that was amazing about it, activism wasn't part of the original plan for the unit. However, the kids came back so fired up after that field trip, that the teachers were like, "Okay. Let's write letters to me [the principal]." Then, from there, I told the teachers, "I don't know what I can do about this." I said, "What if we encourage them to write to Dr. Baird and to the school board?"

If you read Strategy 5, then you know these students led the effort to ban straws from their school district by the following school year. In many schools, teachers may attempt to shield students from disturbing information and/or they may simply teach about problems without engaging students' emotions in empowering ways. Empathy without action competence develops apathetic, overwhelmed, and disengaged citizens.

Conclusion

Schools have a critically important role to play in educating students about, and taking actions that address, issues of sustainability, interrelated and interdependent issues concerning the environment, society, economies, and individual well-being. School leaders model their own expansive ethic of care, awareness, responsibility, and empathy through the issues they draw attention to and the actions they take, or not. Students will experience school as relevant and empowering, or irrelevant and depressing, depending on the degree to which educators provide them opportunities for engaging in powerfully authentic learning and action. Living systems-minded leaders honor the innate agency of children and develop that agency. The next chapter will share ideas for developing partnerships that support expanding these types of learning experiences for everyone.

Leadership Design Challenges

1. **Assess where and how the C.A.R.E. model is operating in your school community.**

   a. **Where do you see an expansive ethic of care expressed?**

   An expansive ethic of care includes concern for today's children, tomorrow's children, and the planet. Review your school's mission and vision statements. What is at the center of your stated circle of concern? If you asked students, families, and community members, what issues would they list as central to their concern? How is your school listening to, and responding to, these issues of concern? How might you expand this work?

   b. **When you consider students' awareness of local and global issues, what makes the list?**

   Goleman and Senge (2014) explained:

   > We must remember that for the first time in human history, children grow up today in the world. By the age of seven or eight, kids are quite aware of the larger environmental and social problems in the world. They can connect the dots. They know these issues will shape the world in which they live. What they are most lacking is a sense that their schools also know and can help them prepare to be able to do something about these issues. (p. 84)

   In what ways does this quote describe, or not, the situation in your school community? Collect and map examples across the curriculum where students have opportunities to express and expand their awareness of local and global challenges. Where are educators missing opportunities to raise awareness? Where might educators better model their own awareness of local and global challenges?

   c. **In what ways are students encouraged to take responsibility and develop action competence?**

   Mogensen and Schnack (2010, p. 62), in their discussion of action competence, explained that

   > . . . the action competence approach points to democratic, participatory and action-oriented teaching—learning that

can help students develop their ability, motivation and desire to play an active role in finding democratic solutions to problems and issues connected to sustainable development . . .

Where are your students developing their sense of efficacy for addressing challenging issues at local and global levels? If your school is like most schools and these opportunities are scarce, then you might consider developing a team of interested teachers to learn together first. Place-based learning resources would be a powerful place to begin. We suggest a book study with one of the following books:

- Vander Ark, T., Liebtag, E., & McClennen, N. (2020). *The Power of Place: Authentic Learning Through Place-Based Education*. ASCD.
- Anderson, S. K. (2017). *Bringing school to life: Place-based education across the curriculum*. Rowman & Littlefield.
- Smith, G. A., & Sobel, D. (2014). *Place-and community-based education in schools*. Routledge.

d. How is empathy modeled and cultivated in your school?
In what ways are you teaching social emotional learning (SEL)? How are adults modeling empathy? To what extent is developing empathy tied to agency? Identify the ways in which students are seeing and developing empathy alongside agency. Intentionally design more opportunities for students to do so as you expand your place-based learning opportunities.

 # Learning From Living Systems-Minded Trailblazers

Abundant Beginnings, an educational collective outside of San Francisco, CA, offers an educational approach grounded in nature, social responsibility, and ancestral wisdom. We are highlighting their work as Trailblazers because theirs is a transformative model that centers themes of social and environmental justice, while also developing leadership and advocacy capacities in young learners. Abundant Beginnings describes themselves as,

> a collectively run, black-led community education and empowerment initiative that is re-imagining how communities can grow learners who think critically, live responsibly, and create meaningful change.
>
> (http://abundantbeginnings.org/)

Children in their care are fully engaged in understanding local social and environmental issues, as well as acting in service to the needs they learn about. The majority of children's time at Abundant Beginnings is spent outdoors, where educators trust and empower children to explore, discover, and learn from their own internal sense of motivation. Developing a deep connection with nature fuels their connections with each other and sets them up for a lifetime of heightened awareness around issues of injustice and oppression. Educators at Abundant Beginnings embrace the reality that children, even at the earliest ages, are deeply attuned to issues of fairness. They believe, in their own words, that

> it is our responsibility and our opportunity as educators and families to engage in dialogue in an explicit and developmentally-appropriate way with our children, about how our world is unfair and how we can use our power to create a better one.
>
> (http://abundantbeginnings.org/)

Abundant Beginnings is modeling an expansive ethic of care and demonstrating how to engage children in honoring their own awareness, taking responsibility for their power, and acting from empowered empathy.

References

Almers, E. (2013). Pathways to action competence for sustainability: Six themes. *The Journal of Environmental Education, 44*(2), 116–127. doi: 10.1080/00958964.2012.719939

AtKisson, A. (1999). *Believing Cassandra: An optimist looks at a pessimist's world*. New York: Chelsea Green Publishing Company.

Bass, L. R. (2020). Black male leaders care too: An introduction to black masculine caring in educational leadership. *Educational Administration Quarterly, 56*(3), 353–395. doi:10.1177/0013161X19840402

Bottery, M. (2014). Leadership, sustainability, and ethics. In C. M. Branson & S. J. Gross (Eds.), *Handbook of ethical educational leadership* (pp. 81–92). New York: Routledge.

Brown, K., Adger, W. N., Devine-Wright, P., Anderies, J. M., Barr, S., Bousquet, F., . . . Quinn, T. (2019). Empathy, place and identity interactions for sustainability. *Global Environmental Change, 56*, 11–17. doi:10.1016/j.gloenvcha.2019.03.003

Chawla, L. (1998). Significant life experiences revisited: A review of research. *Journal of Environmental Education, 29*(3), 11–30.

Chawla, L. (2009). Growing up green: Becoming an agent of care for the natural world. *The Journal of Developmental Processes, 4*(1), 6–23.

Cheang, R., Gillions, A., & Sparkes, E. (2019). Do mindfulness-based interventions increase empathy and compassion in children and adolescents: A systematic review. *Journal of Child and Family Studies, 28*, 1765–1779. doi:10.1007/s10826-019-01413-9

Clayton, S. (2020). Climate anxiety: Psychological responses to climate change. *Journal of Anxiety Disorder, 74*, 1–7. doi:10.1016/j.janxdis.2020.102263

Derr, V. (2020). Diverse perspectives on action for positive social and environmental change. *Environmental Education Research, 26*(2), 219–237. doi:10.1080/13504622.2020.1715925

Eklund, K., Kilpatrick, K.D., Kilgus, S. P., Haider, A., & Eckert, T. (2019). A systematic review of state-level social-emotional learning standards: Implications for practice and research. *School Psychology Review, 47*(3), 316–326. doi:10.17105/spr-2017.0116.V47-3

Ghandour, R. M., Sherman, L. J., Vladutiu, C. J., Ali, M. M., Lynch, S. E., Bitsko, R. H., & Blumberg, S. J. (2019). Prevalence and treatment of depression, anxiety, and conduct problems in US children. *The Journal of Pediatrics, 206*, 256–267. doi:10.1016/j.jpeds.2018.09.021

Goleman, D., & Senge, P. M. (2014). *The triple focus: A new approach to education*. Florence, MA: More Than Sound, LLC.

Jensen, B. B., & Schnack, K. (1997). The action competence approach in environmental education. *Environmental Education Research, 3*(2), 163–178. doi:10.1080/1350462970030205

Kensler, L. A. W., & Uline, C. L. (2017). *Leadership for green schools: Sustainability for our children, our communities, and our planet.* New York: Routledge/Taylor & Francis Group.

Klimek, A., & AtKisson, A. (2016). *Parachuting cats into Borneo: And other lessons from the change café.* White River Junction, VT: Chelsea Green Publishing.

Louis, K. S., Murphy, J., & Smylie, M. (2016). Caring leadership in schools: Findings from exploratory analyses. *Educational Administration Quarterly, 52*(2), 310–348. doi:10.1177/0013161x15627678

Mogensen, F., & Schnack, K. (2010). The action competence approach and the "new" discourses of education for sustainable development, competence and quality criteria. *Environmental Education Research, 16*(1), 59–74. doi:10.1080/13504620903504032

Monroe, M. C., Plate, R. R., Oxarart, A., Bowers, A., & Chaves, W. A. (2019). Identifying effective climate change education strategies: A systematic review of the research. *Environmental Education Research, 25*(6), 791–812. doi:10.1080/13504622.2017.1360842

Starratt, R. J. (2014). The purpose of education. In C. M. Branson & S. J. Gross (Eds.), *Handbook of ethical educational leadership* (pp. 43–61). New York: Routledge.

Taylor, M., & Murray, J. (2020, February). "Overwhelming and terrifying": The rise of climate anxiety. *The Guardian*, pp. 1–4. Retrieved from www.theguardian.com/environment/2020/feb/10/overwhelming-and-terrifying-impact-of-climate-crisis-on-mental-health?CMP=fb_gu&utm_medium=Social&utm_source=Facebook&fbclid=IwAR0LpL0C5CmJ86JTpRf1gk6xbqfvidxfFgIjld5Ry7pmH_6F1Gayk9gZ1Ig%23Echobox=1581327344

8 Partner

Build Caring Partnerships

Living systems-minded school leaders cultivate caring partnerships that have potential to revitalize their schools and their communities. As they embrace whole school sustainability as a means to maximize student learning, they also develop a keen sense of responsibility for their community's well-being and the well-being of the natural world upon which it depends. As school leaders pursue the mutually reinforcing aims of maximizing learning and ensuring community well-being, they are likely to discover powerful partners in reimagining day-to-day school life and in securing future life on Earth.

This strategy outlines various steps living systems-minded school leaders take as they (1) forge organizational relationships by fostering personal ones; (2) allow ideas for action to emerge from across the school community; (3) ensure participants enjoy mutual benefit as a result of partnerships; (4) maintain partnerships; and (5) empower students as partners and change agents.

Forge Organizational Relationships by Fostering Personal Relationships

Living systems-minded school leaders assume the role of lead learner in studying their community's needs, understanding that their local community is characterized by its particular landscape and ecology, as well as by its unique socioeconomic and sociocultural context (Smith & Gruenewald, 2008). As school and/or district leaders engage with their families, community agencies and organizations, local businesses, faith communities, governmental and nongovernment institutions, they begin to know and be known by individual community members. These person-to-person connections establish the foundation for future partnerships. Dr. Baird underscored the power of relationships in building networks for action.

> In my mind, partnerships are just organizational relationships and organizational relationships are built upon personal relationships. Our philosophy has been that everyone we talk to is a potential partner. It's a "yes, and let's see where we go." Oftentimes, we began these relationships without really having a specific purpose for the conversation, not a "you will do this for us, and we'll do this for you", but rather [a conversation] to develop a relationship.

In studying school and community partnerships for sustainability, Wheeler, Guevara, and Smith (2018) similarly proposed that partnerships be understood as dynamic resources, rather than time-specific, transactional arrangements made to address a funding, skill, or personnel need. As living systems-minded school leaders welcome interested community members into two-way dialogue, these discussions nurture mutually beneficial relationships, tapping creativity and talent present within the school community and across the community-at-large.

Allow Ideas for Action to Emerge From Across the School Community

School leaders who acknowledge the ways living systems self-organize to adapt and thrive come to view their school communities as active, living environments of ideas and practical solutions (Clarke, 2012).

These living systems-minded leaders take time to learn how their own systems have responded creatively to changing conditions over time. They may find teachers, principals, custodians, parents, and community members who possess the necessary expertise to participate in developing sustainability-focused learning opportunities as a way to maximize student learning and cultivate a stronger, healthier local communities. In fact, these individuals and groups may already be working to do so.

In many school communities, where whole school sustainability is practiced, Green Teams become a vehicle for bringing champions and experts together to serve as problem solvers and centers of creativity. These Green Teams are often comprised of students, teachers, principals, custodians, parents, and vested and/or potential community partners. As we learned in Chapter 1 of this book, Encinitas Union School District's (EUSD) district Green Team was a starting point for action. Dr. Baird described the genesis of the team.

> The district Green Team was established during my first year in the district. I had already signaled to the community that these issues around sustainability and stewardship were important to me and, hopefully, to the district. We sent out an open invitation to all staff, not just teachers or principals or directors, but all staff, and also to parents and interested community members. We didn't have a lot of thoughts about what the Green Team was going to do at that point. We knew that I was going to be there, not just at that first meeting and not to lead, but as a participant in all meetings. And, our facility director, our food service person, our purchasing agent, and our assistant superintendent were all going to be there. We said, "We are going to be at the table. Come talk to us."

According to Goleman, Bennett, and Barlow (2012), no one person has "the capacity to understand all the ways in which human systems interact with natural systems", and so, ecological intelligence is "inherently *collective*" (p. 7, emphasis in original). In fact, school communities, like all other ecosystems, "come to life through networks of relationships. [They] are ideal places to nurture this new and essential ecological sensibility" (Goleman et al., 2012, p. 7). Several EUSD parent partners have played a vital role in building their community's ecological responsiveness, along with that of other communities up and down stream (Addi-Raccah, Amar, & Ashwa, 2018). A retired parent, with a professional background in wastewater management, joined two other parents to approach the superintendent with an idea for a student-led storm water management program. They piloted what is now called the Storm Water Pollution Prevention Program (SWPPP) in EUSD schools, and then decided to take the program to scale, creating their own company (www.bckprograms. com/). Over the past decade, they have delivered storm water prevention programming at all EUSD schools, as well as with elementary, junior high, and high school students throughout the state. (Figure 0.1. Encinitas Union School District Green Initiatives, beginning on p. 22). With direction from industry specialists, students write their schools' plans, conduct all testing, and share their results at area water conferences. One EUSD principal explained,

> All EUSD schools have a plan. Students work with the facilities department to write the school's plan for the year. When there's a rain event, they test the water runoff and determine what type of pollutants are in the water and then, from there, they develop best management practices. I learned this from my students, that is, what we need do to make sure the water that is running off is clean water. One of the things the students determined is that we have a lot of bird droppings on the roofs of the schools that goes into the storm water. So, we've developed a plan to discourage the birds from perching there.

Likewise, parents have shared their master gardening skills. One parent developed an entire gardening curriculum for her child's school and trained other parents to assist in delivering it. Several parents, who served on the district Green Team, were looking to have an impact on the district as a whole. They volunteered to conduct research and, at a point, Dr. Baird hired them as consultants.

> It wasn't a lot of money, but they formed a nonprofit and became my go-to researchers. They would also go to schools to be a resource to staff on different sustainability-related matters. I'd say, "I need to find the best green cleaning agent out there." Or, "Tell me what sort of integrated pest management is happening and which school districts are implementing this correctly."

EUSD partnerships have emerged from suggestions made by teachers, staff members, parents, principals, the assistant superintendent, and district directors. One important partnership came about through a teacher who knew someone at the Jois Foundation, currently the Sonima Foundation. The Foundation offers health and wellness programs to schools and communities in California, Florida, New York, and Texas via a curriculum that provides children skills to minimize stress, lower incidences of bullying and violence, and improve school attendance and academic performance. Offerings include yoga-based exercises, mindfulness practices, and nutrition education (www.linkedin.com/company/sonima-foundation). The partnership began with several free yoga sessions in a teacher's classroom and eventually extended to regular yoga-based exercise for students at all EUSD schools. In service to communities beyond Encinitas, EUSD provided the foundation a laboratory for developing their health and wellness curriculum, which is now delivered in 89 schools, across the four states just mentioned.

Rather than imposing change from positional power atop the organizational chart, living systems-minded leaders cultivate the conditions in which change happens and creativity emerges from all across the system, generated within networks characterized by high levels of trust and power sharing (Kensler & Uline, 2017). As participants come to understand how their various role(s) connect, or could connect, they begin to see how they might contribute to something larger than themselves. These realizations help them feel valued and empowered to act.

Ensure Community Partners Enjoy Mutual Benefit

All living things are connected, directly and/or indirectly, relying upon dense networks of relationships for their well-being (Capra, 2002). Ecologists document the various ways species diversity, partnerships, and networks help communities withstand and recover from major disturbances, demonstrating resilience (Peterson, Allen, & Holling, 1998). Traditional approaches to school/community relations have often failed to grasp the importance of interdependent relationships, viewing partnerships more narrowly as serving only school-related outcomes, such as academic achievement and parental involvement (Green, 2018). Increasingly, scholars of educational leadership have emphasized community-wide development as an alternative vision for school/community partnerships, focusing more broadly on improving neighborhood, as well as school, outcomes (Crowson & Boyd, 2001; Green, 2015; Keith, 1996; Khalifa, 2012). In this broader context, schools and school districts become "enmeshed with other community agencies in an interconnected landscape of supports for the well-being of students and learners" (Goldring & Hausman, 2001, p. 195).

Living systems-minded leaders go a step further, extending this social landscape to include the natural world. They understand the outcomes of community partnerships as including the well-being of children, the well-being of their community and communities up and down stream, and the well-being of planet Earth. As Capra (2009) explained,

> This is the profound lesson we need to learn from nature. The way to sustain life is to build and nurture community. A sustainable human community interacts with other communities—human and nonhuman—in ways that enable them to live and develop according to their nature.
>
> (p. 2)

Dr. Baird reflected on the interdependent and reciprocal relationships the district forged with various community partners over time.

> And it usually starts with, "What are our mutual goals?" You've got to start with trust and a common belief system. But, from that trust and common belief system, then you have to arrive at an understanding of "what do you need from this partnership, and what can I help you with?" The basis of an organizational relationship might be just one thing. But, as you grow together in relationship, you start planning together. You start talking about how we can now do different things together, as a team.

The relationship between the EUSD and The Ecology Center (www.theecologycenter.org/) provides a robust example of a mutual partnership, one that supports school district, community partner, and community-wide needs. The Ecology Center is a 28-acre organic farm, farm stand, and ecological education center located in San Juan Capistrano, California. Over the past decade, this nonprofit organization has worked with over 100 schools and trained thousands of community members. The organization's website shares their beliefs that "everyone should have access to the tools, knowledge, and skills that promote healthy communities in the 21st century" (www.theecologycenter.org/). Dr. Baird described the search for an appropriate partner to run the school district's organic farm.

> Once we developed Farm Lab, we realized we wanted to run a farm there, but we weren't farmers. So, then we thought, "Let's find a partner." We actually explored five different potential partnerships before we landed where we are now. With some, we were in sync around our beliefs and there was high trust, but the partner didn't have the ability to deliver. With others, we realized their goal was not about serving students and the community, but rather, about making money and controlling the land. Then we learned about the work of The Ecology Center, their outreach to schools, and their community-learning center (See Figure 8.1). We got in contact with them, developed that relationship, and they then expressed a desire to come on campus.

Julie Burton, Coordinator for Innovation and Development at the EUSD Farm Lab, elaborated,

> The Ecology Center is our farming partner. We pay them a very reduced rate for farming the gardens that provide our students a living laboratory and supply our schools' food service program. They are also building out one side of the DREAMS Campus property into a full community education space. Community members will be able to come, learn how to grow their own food, and minimize their footprint in any way you can imagine. They'll learn how to create their own little ecosystem at home.

During the COVID-19-related school closures in spring 2020, school salad bars across the district were, of course, closed. During this time, Farm Lab staff, in partnership with the Ecology Center farmer, continued to harvest all produce at the farm. Food not utilized by Child Nutrition was donated to the community, with free lunches distributed through July and families who were experiencing food insecurities receiving free organic produce.

Living systems-minded school leaders abandon limited notions of school/community relations, in favor of developing more interdependent partnerships. In doing so, they help create networks of action for realizing genuine social change (Gough, 2005). In this reinvented context, students,

Figure 8.1 The Ecology Center Outreach Bus visits schools in the area.

teachers, principals, parents, and community partners work together to create healthier communities and move the sustainability agenda forward (Gough, 2005).

Maintain Partnerships

The capacity to solidify and extend emerging green initiatives rests in leaders' capacity to not only develop, but also maintain, dense and nimble networks of relationships across the district and out into the community. Living systems-minded leaders serve as the guardians of these connections, acting as facilitator, negotiator, and champion. They celebrate and support ongoing actions, communicating the cost-saving and value-added outcomes for schools, students, and community members. Knowledge of these benefits provides impetus to keep partners energized and engaged. Dr. Baird underscored the importance of authentic involvement in efforts that make a difference. "Give your committees some teeth. Make them understand that they're there for a reason. If you bring people to the table, empower them to do something, and then validate the work that happens."

Living systems-minded leaders continuously scan their community, connecting various pockets of innovative work. For EUSD, the Farm Lab became a nexus for such innovation, connecting key community institutions within the immediate neighborhood, including San Diego's botanic garden, an organic local farm committed to sustainable agriculture and food justice, a local history museum, the YMCA, a retirement community, and the school district. The leaders of these institutions began with exploratory conversations, talking and thinking about things they might do together. Over a number of years, the personal connections developed into a more formal partnership known as the E3 (Encinitas Environmental Educational) Collaborative. Dr. Baird spoke to the evolution of the partnership, reflecting, "We have been the connective tissue that holds the collaborative together, because our students and the school district work so closely with all of these members." In order to maintain the momentum of their various joint projects, the E3 Collaborative established a formal nonprofit. Partners signed a memorandum of understanding (MOU), embracing the shared goals of

> preserving and encouraging access to nature; supporting sustainability education and the health and well-being of people of all ages, backgrounds and

179

abilities; developing multigenerational learning programs focused on agriculture, horticulture, nutrition, science, sustainability, community building, and the local history of Encinitas.

(http://encinitascollaborative.org/)

Dr. Baird explained, "It's a form of communication, and the messaging around it is clear. This transcends the personal. The personal is necessary, but the organizational relationship lives on after the people leave."

The school district has formalized partnership efforts in other ways, assigning these responsibilities to a district-level leader who supports the efforts of the superintendent, assistant superintendent, principals, and teachers in interacting with outside entities and helping to cultivate, develop, and support sustainability-related partnerships. Dr. Baird summarized, "If something is important, you put it in your formal documents, you put it in your plans, you put it in your budget. You put adequate resources behind it."

Empower Students as Partners and Change Agents

To the extent that adults trust children to actively wrestle with the critical concerns that challenge our world, these public issues gain personal meaning within their lives (Chawla & Cushing, 2007). As students engage over extended periods of time, they not only learn necessary knowledge and skills, but also achieve valued goals (Chawla & Cushing, 2007, p. 441). According to Chawla (2007), these sorts of primary, "full-bodied" experiences motivate children "to protect the places they love and [build] alliances and competencies to do so" (p. 153). Venturing forth from school to engage in real-world, problem-based learning, children build a sense of individual and collective agency as they undertake meaningful endeavors and succeed (Chawla, 2009, p. 16). Through these experiences, students discover they are wholly capable agents, prepared to make important things happen in the world (Kensler & Uline, 2017). In a group interview, sixth-grade students at an EUSD elementary school, told the story of their Ballot Bin initiative, aimed at removing trash from their storm water run-off. The student's efforts began the previous year in the context of a fifth-grade Civics project. Their prototype for the Ballot Bins is pictured in Figure 8.2.

Figure 8.2 The students created a prototype for their Ballot Bins.

Student 1:   We were looking at Senate bills for Civics class. We wanted
              to do an environmental one. We picked Senate Bill 54, which
              is about eliminating single-use plastics. We were looking for
              ways that we could implement that in our school. It's not the
              whole world. We have to start with our school.

Student 2: The storm water pollution was rising at our school. Our Storm Water Pollution Prevention Plan (SWPPP) team tested the storm water and found many micro plastics and big pieces of trash. We researched and found the idea of so-called 'Ballot Bins' from a city in England where they created these bins to remove cigarette butts from their streets and sidewalks. There would be topics to vote on with two choices, such as 'Pick your favorite soccer team.' People would vote by putting their cigarette butt in one side of the bin, which had a glass divider to keep votes separated.

Student 3: Although there may not be any reward, it's a fun way of picking up trash. It's not that fun to pick up trash, no matter how environmentally friendly you may be. But we all love voting, so why not combine the two into one?

Student 4: We got Home Depot to sponsor us and now the bins are everywhere. Home Depot provided enough materials for 30 Ballot Bins and helped us build them. We kept 10 of them for our school and the school district, and then we gave some to a town closer to the border [with Mexico].

Student 5: We had teams make posters and go into all the classrooms to explain about the Ballot Bins. As soon as we introduced them, everyone was really excited about picking up trash and voting with it in the bins.

Teacher 1: They taught the school. It was a hundred percent them, with very little adult direction.

Student 3: We presented on our project at The Zero Waste Symposium [a regional forum]. The whole fifth grade went as a field trip. We listened to different speakers talk about how they prevent plastic from going into the ocean. We had a hundred kids in a conference with lots of adults and high school and college students. I think they were surprised to see elementary kids listening and taking notes.

Student 2: It was cool that everyone shared their own ways of helping. If we're all trying to solve a problem the same way, we may only get a fraction of it solved. If we all find ways that are different from one another's, then we can each tackle a different side of the spectrum. It may not be possible in my lifetime, but maybe I can pass it down for the generations to come, and we can start destroying the great Pacific garbage patches and save our world.

The fifth-grade Ballot Bin initiative provided these fifth-grade learners the opportunity to partner with each other, their teachers, and other committed adults throughout their community and their region. Such authentically inter-dependent learning experiences help students develop the skills and competencies necessary for partnership, participation, and action (Henderson & Tilbury, 2004). Project outcomes also reinforced their sense of optimism and agency. When students engage in such nature-based, in-community learning, they build a strong foundation for their current and future work as stewards and change agents, engaging in environmentally focused endeavors and sharing their experiences with interested adults (Malone, 2013).

Conclusion

As discussed within the Introduction of this book, schools and school districts live within larger social, economic, and ecological systems with which they are interdependent. Living systems-minded school leaders explore these local-to-global interdependencies as opportunities for broadening their spheres of influence. They collaborate with like-minded partners, within their school communities and across their communities-at-large, distributing and/or sharing their power by empowering others. As they do so, they co-create active networks of relationships, capable of accomplishing their sustainability-related goals (Wheatley, 1999). These caring partnerships, and the authentic relationships upon which they depend, represent a primary source of energy for creating vibrant, flourishing, sustainable schools and communities. And, as their students learn, by example and through active participation, they also develop their identities as builders and keepers of community (Niemi & Junn, 1998).

 Leadership Design Challenges

1. **Create a blueprint for a district-level Green Team.**
   Begin with the obvious key players, including teachers, principal(s), parent leaders, and district-level department administrators (e.g., nutrition services, curriculum, purchasing, maintenance and operations, planning and construction, etc.). Scan the larger community for local nonprofit organizations involved in environmental and/or social issues. Identify appropriate local governmental organization representatives.

Explore the missions of various local businesses and private foundations for potential matches. Consider faith communities who might be interested to participate. Organize your list according to full and ad hoc members. Invite participants to the table for an exploratory conversation, asking the question, "What might we do together to create a more vibrant, flourishing, sustainable school/school district and community?" Listen, record responses, and follow up.

2. **Prepare a 10- to 15-minute presentation on the power of partnerships in creating sustainable schools and communities**.

Describe various ways community members can participate in, and benefit from, the creation of greener, more sustainable schools. Touch on the inherent learning opportunities for students, teachers, parents, and the community-at-large. Help your audience understand how, together, you can ensure that sustainability- and nature-based learning will not be another passing fad in your school/school district, and will, instead, continue to serve both school and community outcomes into the future.

3. **Design a Community Partner Appreciation Day**.

Create an opportunity for your community partners to see the fruit of their labors. You might identify a yearly theme to highlight in accordance with your school and/or district learning goals. Choose several schools where these themes are alive and thriving through sustainability- and nature-based learning. Invite your community partners to board bus(es) to each school. As you drive between schools, provide background on what they are apt to see. Explore each school together. Following the tour, serve your partners a meal or snack, accompanied by guided conversation around questions such as, "How do you understand the needs of students in the 21st century and beyond? In what ways does what you saw today match with your expectations?" Close with recognition of each partner and acknowledgment of the partner's specific contribution.

 Learning From Living Systems-Minded Trailblazers

In the June 2018 issue of *Green Schools Catalyst Quarterly**, Dr. Michael Salvatore, Superintendent of Long Branch Public Schools (a mid-sized urban district, located along the eastern corridor of New Jersey's coastline), described the school district's efforts to create a culture of sustainability within their school community. At the start of Dr. Salvatore's tenure in 2011, district revenues were in decline, even as student enrollment grew to an all-time high, with a corresponding expansion of students' unique learning needs. In answer to the districts' challenges, Dr. Salvatore went about "cultivating a mindset interwoven with sustainable practices in every school, department, discipline, and function" (Salvatore, 2018, p. 94). Through implementation of a three-phased effort, the district improved their energy efficiency (with energy savings exceeding $360,000 in the first budget cycle); launched a high-quality medical care center for employees (which not only reduced costs, but also synthesized data to coordinate care and offered a host of wellness services including chiropractic, physical therapy, acupuncture, dietary-nutritional services, as well as yoga for employees and their dependents); and expanded a rigorous sustainability-focused, project- and problem-based learning program for all students. Dr. Salvatore described "[e]xtremely ambitious 'green teams' . . . assembled in our schools [who] quickly identified partnership organizations to enhance sustainable learning experiences, both during the school day and beyond" (Salvatore, 2018, p. 95).

Dr. Salvatore reported an ever-expanding network of "partners in sustainability" which includes, but are not limited to, Sustainable Jersey for Schools; Family Success Center LB; Surfers Environmental Alliance; New Jersey Watershed Organization; Monmouth Medical Center; Monmouth University; Clean Ocean Action; Sodexo Food Services; Jersey Central Power & Light; Long Branch Environmental Commission; Rutgers University Master Gardeners; B.U.L.B.S. Garden Club; New Jersey Natural Gas; New Jersey Audubon Society; National Wildlife Federation; Monmouth County Food Bank; Long Branch Food Banks; Long Branch Ministerium; and the City of Long Branch (Salvatore, 2018, p. 95).

He described sustainability initiatives as "a scaffolded process beginning with common practices, such as creating school gardens, and scaling up to more innovative learning opportunities, such as designing a home with sustainable materials in our engineering classes" (Salvatore, 2018, p. 96). He elaborated.

> Our collaborative nature manifested a 'once in a lifetime' opportunity for our children, faculty, project sponsors, and community when we constructed a house, from foundation to roof, for a family in our city. This conceptualization originated in a high school classroom during a partnership meeting. The students' enthusiasm was so contagious that a local city official and school administrators got together and identified a parcel of land. Habitat for Humanity guided us through the process and expanded the effort throughout the city, from toddlers to senior citizens. Everyone who heard of the project was motivated to join in.
>
> (Salvatore, 2018, p. 96)

Dr. Michael Salvatore and Long Branch Public Schools provides a powerful example of what can happen when living systems-minded school leaders, in collaboration with like-minded community partners, set out to create vibrant, flourishing, sustainable school communities. The sky really is the only limit on what is possible!

Green Schools Catalyst Quarterly (https://catalyst. greenschoolsnationalnetwork.org/gscatalyst/Store.action) is a peer-reviewed digital magazine that highlights evidenced-based practices for replication in green, healthy, sustainable schools. Along with articles by researchers and experts in the field, the publication shares powerful stories from Living Systems-Minded Trailblazers throughout the United States.

References

Addi-Raccah, A., Amar, J., & Ashwa, Y. (2018). Schools' influence on their environment: The parents' perspective. *Educational Management, Administration & Leadership, 46*(5), 782–799.

Capra, F. (2002). *Hidden connections*. New York: Doubleday.

Capra, F. (2009). *The new facts of life*. Center for Ecoliteracy. Retrieved from Center for Ecoliteracy website: www.ecoliteracy.org/essays/new-facts-life

Chawla, L. (2007). Childhood experiences associated with care for the natural world: A theoretical framework for empirical results. *Children, Youth, and Environments, 17*, 144–170.

Chawla, L. (2009). Growing up green: Becoming an agent of care for the natural world. *Journal of Developmental Processes, 4*, 6–23.

Chawla, L., & Cushing, D. F. (2007). Education for strategic environmental behavior. *Environmental Education Research, 13*, 437–452.

Clarke, P. (2012). *Education for sustainability.* Abingdon, Oxon: Routledge/ Taylor and Francis Group.

Crowson, R. L., & Boyd, W. L. (2001). The new role of community develop- ment in educational reform. *Peabody Journal of Education, 76*, 9–29. doi:10.1207/S15327930pje7602_2

Goldring, E. B., & Hausman, C. (2001). Civic capacity and school principals: The missing links for community development. In R. Crowson (Ed.), *Community development and school reform* (pp. 193–209). London: Elsevier.

Goleman, D., Bennett, L., & Barlow, Z. (2012). *Ecolierate: How educators are cultivating emotional, social, and ecological intelligence.* San Francisco, CA: Joseey Bass.

Gough, A. (2005). Sustainable schools: Renovating educational processes. *Applied Environmental Education & Communication, 4*(2), 339–351.

Green, T. L. (2015). Leading for urban school reform and community devel- opment. *Educational Administration Quarterly, 51*(5), 679–711.

Green, T. L. (2018). School as community, community as school: Examining principal leadership for urban school reform and community develop- ment. *Education and Urban Society, 50*(2), 111–135.

Henderson, K., & Tilbury, D. (2004). *Whole-school approaches to sustain- ability: An international review of whole-school sustainability programs.* Canberra, Australia: Australian Research Institute in Education for Sustainability.

Keith, N. (1996). Can urban school reform and community development be joined? The potential of community schools. *Education and Urban Society, 28*, 237–268.

Kensler, L. A. W., & Uline, C. L. (2017). *Leadership for green schools: Sustainability for our children, our communities, and our planet.* New York: Routledge/Taylor and Francis Group.

Khalifa, M. (2012). A "re"-new-"ed" paradigm in successful urban school leadership: Principal as community leader. *Educational Administration Quarterly, 48*, 424–467.

Malone, K. (2013). The future lies in our hands: Children as researchers and environmental change agents in designing a child-friendly neighborhood. *Local Environment, 18*, 372–395.

Niemi, R. G., & Junn, J. (1998). *Civic education: What makes students learn*. New Haven: Yale University Press.

Peterson, G., Allen, C. R., & Holling, C. S. (1998). Ecological resilience, biodiversity, and scale. *Ecosystems, 1*, 6–18.

Salvatore, M. (2018, June). The future you want now: Creating a culture of sustainability in a New Jersey school district. *Green Schools Catalyst Quarterly*, 94–97.

Smith, G., & Gruenewald, D. (Eds.). (2008). *Place-based education in the global age: Local diversity*. New York, NY: Lawrence Erlbaum Associates.

Wheatley, M. J. (1999). *Leadership and the new science: Discovering order in a chaotic world* (3rd ed.). San Francisco, CA: Berrett-Koehler.

Wheeler, L., Guevara, J. R., & Smith, J. A. (2018). School-community learning partnerships for sustainability: Recommended best practice and reality. *International Review of Education, 64*, 313–337.

STRATEGY

9

Start Small

Start Small and Stay
Anchored in a Vision
of Vibrant, Flourishing,
Sustainable Schools

Co-authored by Joy Miller

Living systems-minded school leaders facilitate the development of shared visions for sustainability. These visions are expansive, transformative, and motivating. Through articulation and implementation of such visions, living systems-minded leaders reveal, direct, and strengthen the interdependent connections between each action they and members take to realize their vision of vibrant, flourishing, sustainable schools.

The vision of a social system connects its members to a unifying purpose; it creates the momentum to move forward in reimagining potential and in plans for action. Vision provides direction, but more than this, it defines the trajectory that determines a future destination. Schools that intentionally place their statements of vision and mission within each layer of the organization reap the benefit of cultivating a shared focus that guides the system through its seasons of growth and adjustment. We recognize, though, that not every leader will do this. There is great variance in how schools understand and prioritize their purpose (Stemler, Bebell, & Sonnabend, 2011). Yet, when living systems-minded leaders confidently cast a vision that speaks to a place beyond current conditions, and then pair a sense of purpose with action steps, they find a catalyst for deep and meaningful change.

Research shows that when vision, mission, and purpose are clearly aligned and communicated, schools function more effectively (Hawley & Rollie, 2002; Pekarsky, 2007; Sammons, Hillman, & Mortimore, 1995). While only a few trailblazing schools have begun to explicitly embed themes of sustainability into their statements of purpose and vision (Kensler & Uline, 2017), this step is critical in encouraging a lasting shift in educational landscapes. Positional school leaders have the opportunity to support the conditions that nurture interest in, and movement towards, green priorities by communicating and connecting them to the core of their learning organization. Consider what this looks like in Encinitas Union School District (EUSD). Dr. Baird points out, referencing EUSD's four pillars that we described in detail in the Introduction,

> Everybody in our district knows environmental stewardship is one of our core pillars. They can talk about things we do that support that pillar. They know health and wellness is one of our four pillars. Those two go together and in a lot of different ways, but they know that.

Leaders can use vision and mission to spark dialogue and to frame aspirations in such a way that new behaviors and language begin to naturally spread into school culture and daily practice. As the community internalizes shared intentions, in time, schools will reshape themselves and become the living expression of this clearly communicated, courageous vision. In this chapter, we speak to you, the reader, directly. You are the trailblazers that are leading in this 21st century. In the words of Jane Goodall,

> You cannot get through a single day without having an impact on the world around you. What you do makes a difference, and you have to decide what kind of difference you want to make.

Start Small

While the picture of a sustainability-conscious learning community is expansive and its vision points to significant, inspiring outcomes, it is important to start with small steps that support a natural process of change. Emerging values will form and unfold in incremental phases, and so leadership teams will want to be both action-oriented and patient. Research affirms that organizational change includes an emotional component and usually a mindset shift (Burke, 2018), and so pushing faster and further than people are prepared to move is counterproductive to the sustained growth living systems-minded leaders want. Granted, those who sense the immense responsibility they carry in leadership may long for immediate, revolutionary change, but school systems tend to be evolutionary, and thus, they adapt slowly and in stages.

For this reason, this last strategy will offer some small steps that leaders can take to move their learning community towards greater stewardship and sustainability. Wherever your school is today, you are positioned to move it forward. School leaders have the challenge and privilege of preparing the path for teams to practice what has not been done, or even thought of, before. As we discuss practical ways to begin and continue change, consider the process of cultivating and growing a garden as a picture for this process. Like planting a garden, implementing sustainability in schools happens in small steps that foster well-being and inspire growth in ways that are both guided and natural.

Notice the parallels:

1. **Prepare the soil.** The foundational work is done behind the scenes and below the surface in order to lay the groundwork for change. In educational systems, leaders tend to the culture and climate as they cast vision and build capacity in others.

2. **Select and plant the seeds.** When nurturing conditions have formed, it is time to decide which efforts to sow. Leaders consider current interests and strengths among their teams and choose a project that connects with the energy at school and in the community.

3. **Nourish new growth.** Once in place, as the initiative takes root and shows signs of flourishing, changes are supported by connecting to additional resources. At this stage, projects are visible, and leaders seek ways to build up the work that has begun.

4. **Harvest success.** The season of outcomes reveal opportunities to enjoy results, share the rewards, and collect new seeds. Here, the leadership team spreads successes throughout the school community and continues the cycle of change by collecting and planting new seeds.

Prepare the Soil

Gardeners know the importance of preparing the soil before they plant seeds, and education leaders will need to prepare school conditions to embrace new work they want to do. Well before the harvest of substance arrives, the gardener has envisioned possibility and knows of places to plant. This is the visionary leader who stands within their school, seeing potential and holding some of the initial seeds of transformation. As in a garden, these seeds, good and strong as they are, cannot simply be scattered over the ground in hopes that something will take root and thrive. In the garden, soil must be able to receive the seed. In the school, this soil is the culture.

There is no other dynamic that will affect the success of sustainability efforts more than the culture. The school community's shared beliefs, assumptions, and values will be what determines whether initiatives thrive or languish, and so after vision, here is where we start. School leaders hold the positional power to push mandates and require participation from members, but demands coming top-down or outside-in do not align well with the way living systems change (Kensler & Uline, 2017). Transformational leaders are not looking for mechanical compliance, but rather authentic buy-in and holistic ownership among their teams. Ultimately, the goal is to support widespread, systematic change that will not dwindle and die off, but will last even long beyond a leader's tenure. With this in mind, change agents will want to invest time preparing people to step forward so that system members will be able to keep moving by their own free choice.

The desire for school redesign and reform should sprout from within the system's various levels so that sustainability values are nested within its core beliefs. When it does, ownership and commitment are not shallow, but are set deep within the layers of school culture. School leaders can prepare their communities for change by distributing leadership in such a way that networks and partnerships, built on fairness, dignity, and diversity, become the natural expression of the organization. Healthy networks enable both the individual and the collective whole to thrive; dense interconnection supports well-being in social systems just as it does in natural systems. Ecology itself shows that communities built on diverse partnerships and networks endure change better and are more able to adapt and flourish afterwards (Kensler & Uline, 2017; Peterson, Allen, & Holling, 1998). In the educational context, this looks like teams and larger learning communities joining together, drawing upon individual strengths and diverse experiences, in order to advance a common goal.

Networks of support are necessary in all major change initiatives, and they are particularly crucial for schools moving towards a living-systems approach to stewardship. A shift this substantial cannot thrive in isolation, nor can it depend upon the efforts of a few disconnected individuals. Living systems, such as schools, flourish with interconnection and so, by forming trusting networks of teams, leaders plant healthy collaboration into their work. Networks encourage school communities to consider innovative ideas, and they invite fresh insight into conversations. Networks empower many different voices to contribute, and they promote wellness by bringing a variety of perspectives into the efforts.

School leaders can promote a culture and climate of collaborative teamwork by modeling trust and transparency. Trust, when it functions through democratic principles, fosters healthy social systems (Kensler, Caskie, Barber, & White, 2009; Meier, 2002), and trust among members is necessary to protect organizational change from failing (Daly, Moolenaar, Bolivar, & Burke, 2010; Louis, 2007). A trusting, transparent culture produces a climate open to innovation and accepting of positive risk-taking. When learning networks know they can try new things and risk making a mistake, the freedom to grow and change becomes a school value. Leaders who openly display trust in their teams strengthen members to take on greater responsibility and step beyond their comfort zones. These conditions foster well-being in a learning community, and they infuse members with the energy needed to move towards a greater vision for the school.

Reflecting on your system's culture, are there roots from the past that may limit future growth? School culture runs deep, like the soil of the figurative garden; its composition cannot be fully seen from a disconnected vantage point, far above ground level. Leaders will want to get down into the dirt and be willing to dig into what may promote or what may block the seeds of change from settling in. Are there mindsets, fears, beliefs that will need to be uprooted through the work? Is the culture connected? Trusting? Transparent? Innovative? Risk-taking? If not, what can the leadership team do to reset the soil and build new capacity?

Select and Plant the Seeds

When the soil of culture shows pockets of readiness, the green team will be able to select and plant seeds of change in the form of projects and initiatives. At this initial sowing stage, visionaries have ensured that school culture is receptive and can begin leading the way towards a more sustainable school where individuals embrace the value in connecting their community to the natural world. While there could be any number of good places to start, the leader who is committed to long-term, lasting results will consider where the community's energy is most evident. As living systems-minded leaders consider options, they will want to learn what those around them value most.

There are already system members, adults and children, with an interest in green practices and a passion for making a difference, who have great ideas and are ready to act. It is the leader's job to identify what people wants to do, facilitate collaboration, and connect the resources needed to start a project that the community will appreciate. In your school system, which ideas continue to resurface in conversations? Which area would be a simple, natural starting place that would create encouraging, measurable results and be meaningful to the community? The green team at Encinitas had a number of ideas and options for initial projects, but leaders recognized that they could not do it all right away. Thinking back, Dr. Baird shares his thoughts on where to begin.

> Most people start, I always say in the three G's. I would say start with the green team, start with a garden, and start with garbage or waste reduction. Those are the three easiest areas to get your feet wet. We did exactly that.

As leaders consider their launch, they want to think of a small step that will set their systems up for success. Learning communities may resemble each other in many ways, but each is unique in its specific makeup of people, place, and purpose.

Reflecting on your community's characteristics, what would capture attention, invite collaboration, and contribute in the most resonating way?

The first plants in the garden provide a focal point and a topic for discussing additional work that could be done. Leaders know the seeds they sow are important, and so they carefully consider what should take root first. As you evaluate options, what project would be most natural to plant? What compliments the current environment? What would add to the conversation? Ultimately, what is the best fit for the community?

Nourish New Growth

In a garden, the groundwork of preparing the soil and selecting and planting the seeds will bring about new growth. Growing seeds will break through the ground and shoot up into a creation that no longer resembles its modest beginning. Just as a gardener nourishes the sprouting plants, school leaders will find resources to supplement the school's emerging and maturing efforts. This is when leaders will look for external sources, such as regional and national connections, conferences, publications, or programs to further support the growth of the system.

To begin, consider which affiliations, materials, or experiences may best align with your project needs and school interests. The resources provided in *Leadership for Green Schools* have been expanded and updated in the list below:

Auburn University's *Leading Green Schools* course is an online, ten-hour, self-paced course developed for school leaders and their support teams (https://online.catalog.auburn.edu/courses/leadgreen-live). Those who want to implement a whole school sustainability approach will find in this course ten modules of content that will support and guide their efforts. Along with educational materials and engaging, interactive content, *Leading Green Schools* abounds with insights from practitioners and relevant readings that encourage deep reflection and learning. Topics such as: schools and energy, school buildings as teaching tools, culture and well-being, family engagement, equity, vision and mission, and more are detailed with practical examples, videos, research, and opportunities to apply new learning. By the end of the course, participants will have created a comprehensive, professional sustainability plan that can be presented, shared, and implemented in schools.

We mentioned the *U.S. Department of Education's Green Ribbon School* award program (www2.ed.gov/programs/green-ribbon-schools/index.html) in the Introduction. Participating states require an application that can be a powerful self-study for schools and guide transformative processes for those schools still early in their journey.

The *U.S. Green Building Council's Center for Green Schools* (USGBC, The Center) at centerforgreenschools.org provides introductory materials, case studies, research, action guides, and many other resources and programs. For example:

- The *Green Classroom Professional Certificate* is a two-hour online course that provides a great place to start with educator professional development.

- *Green Strides* (accessed via the Center website or greenstrides.org) is a web portal originally developed by the U.S Department of Education to support schools seeking to earn the ED-GRS award. It offers webinars, free tools, key term searches, and a regular newsletter.

- *Learning Lab*, along with *EcoRise Youth Innovations*, *Representaciones Inteligencia Sustentable*, and The Center, is a partnership that provides English and Spanish project-based curriculum aligned to standards. Created by educators for educators, integrating sustainability into the curriculum can be found at https://learninglab.usgbc.org.

- *Professional Networks* across the nation use The Center website to stay in contact with other sustainability minded school leaders and districts. Together, these groups share success, challenges, and learning within the community.

- The *Green School Conferences & Expo* offers an annual opportunity to meet with other administrators, teachers, sustainability advocates, policy makers, and industry leaders. To connect and take the next step in green school practices, this event is abundant in resources.

- *Green Apple Day of Service* (access via The Center website or http://greenapple.org) highlights volunteer work that focuses on practical efforts to green schools across the country. Project ideas, check lists, social media materials, and fundraising information can be found here.

- The following organizations also offer excellent resources for nourishing efforts towards greening your school:

- *Earth Force* (earthforce.org) is a site providing resources and ideas for teacher training and student projects, including PBL's for virtual learners, which tap into local environmental needs. By taking a solution-focused approach, this organization provides frameworks for project implementation and student focused programming.

- The *Center for Ecoliteracy* (www.ecoliteracy.org) offers extensive resources, initiatives, and professional development related to sustainability, system change, and ecological education.

- *Green Schoolyards America* (www.greenschoolyards.org/) is an organization that offers research, policy, and support related to transforming asphalt schoolyards into more engaging, enriching, and enlivening spaces for children's play and learning. They are also a co-founder of the National COVID-19 Outdoor Learning Initiative that supports educators' efforts to effectively open schools for learning during a pandemic. Their work focuses on the methods, opportunities, and benefits of more intentionally and creatively using outdoor spaces for learning.

- The *Cloud Institute for Sustainability Education* (www.cloudinstitute. org) provides consulting, professional development, and innovative curriculum design related to sustainability education through a whole systems approach.

- The *Green Schools Alliance* (www.greenschoolalliance.org) is an international coalition of green-minded schools and leaders who have access to extensive resources and the ability to network. Find information about their *Green Cup Challenge* on their website. This challenge is a competition for reducing energy use and improving recycling/waste reduction programs.

- The *Green Schools National Network* (www.greenschoolsnational network.org) offers an excellent one-pager for guiding school greening efforts. In addition, there is an extensive list of green school networks listed by state to help connect local leaders. In addition, they offer these resources:

  - *Catalyst Schools and Districts*. The vision for the Catalyst Network, comprised of schools and school districts across the country, includes leveraging the triple bottom line to prepare students to co-create a sustainable future while increasing student achievement; improving the health and well-being of students and staff; and decreasing operational costs and the school's/district's ecological footprint.

  - *Green Schools Catalyst Quarterly*. Visiting the Green Schools National Network provides access to this high-quality digital journal focusing on sustainability and health in K–12 schools. While research based and peer reviewed, this publication speaks to practitioners by addressing timely, theme-based topics that will resonate with school leaders.

- Finally, the *Waters Foundation* (www.waterscenterst.org), although not focused on sustainability-related education, offers excellent resources for teaching and learning-systems thinking. The array of systems thinking tools they introduce, define, and model for all grade levels are readily applied to issues of sustainability.

Nourishing the growth of projects nested within an expansive vision for change strengthens current efforts and also opens space for future expansion. The work here primes the school environment to embrace bolder, more involved aspects of sustainability as people experience and see results in the first small steps. Visible success is, itself, a nutrient that feeds a group's momentum to continue towards longer-range goals. With this in mind, reflect upon your community's needs. Do you have adequate access to external resources? What source of nourishment is your system receiving now? Where does the system need additional support to grow? In what way could your team become a resource to others?

Harvest Successes

The gardeners who devote their focus and energy into preparation, planting, and care will come into times of reaping the benefits of their persistent work. In tangible ways, the plants will produce, and gardeners will recognize that conditions are right to shift into harvest; the time when desirable results are collected and shared. Similarly, as school sustainability projects show signs of maturing, and as initiatives create transformative outcomes, the community can celebrate, partake, and even replant to promote new progress.

Taking small steps towards sustainability is an effort worth celebrating. In fact, not only should school leaders harvest their system's successes as they come, but they should actually plan for wins along the way. A recognized leader in organization change, Dr. John Kotter of Harvard University, explained that it is risky to embark on long-term change efforts without planning for "visible, unambiguous success as soon as possible" (1996, n.p.). People need to see some clear rewards for their efforts early, so that energy is replenished and their commitment is further forged to the vision before them. It is human nature to connect to what is healthy and thriving, and so, by showing others an effort's current success, supporters solidify their participation and late adopters take notice.

For these reasons, consider weaving celebration into your culture's pattern of practice. Success is not a single point that marks completion, and so teams need not postpone promoting their accomplishments till the far-off future. In practice, progress and growth are rarely linear; therefore, leaders can celebrate change that comes in cycles, in overlapping movements that authentically reflect the natural rhythm of a project in motion. Kotter (1996) pointed out that, in the corporate world, a long-term change initiative needs to show results at the 14- to 26-month marker in order to ensure a company will fulfill its goal. In the school setting, this time frame likely needs to shorten further, especially because children serve as key participants. Understanding how important this is, leaders can not only look for signs of early success, but they can also strategically plan them. In one of EUSD's first efforts, Dr. Baird shared the good results that come when leaders think about how to celebrate success early.

> I was looking for a quick win to say, "Hey, the green team just saved $40,000 for this district." We did that in the very first couple of months just by looking at where the trash was going. Then we came back with, "Okay, where are we generating most of our trash?" . . . Kids now got involved in some of this work. We realized, at one school site, we cut 83% of the lunchtime waste when we utilized effective recycling and waste reduction education efforts. . . . That model led to us to say, "This works. This can work at other campuses," and it spread to other campuses.

As seen with EUSD, a visible win naturally supplies the energy needed to expand projects and permeate efforts across the district. Living systems-minded leaders set their system up to succeed early, and openly communicate their results in ways that catch attention, thus sparking widespread interest and harvesting seeds of inspiration to use in the future.

Harvesting seeds of success also enables living systems-minded school leaders to use transformative outcomes as a way to promote new growth and build capacity in people, thus continuing cycles of change. Dr. Baird modeled how this looks in practice. He explained,

> It doesn't matter whether you're a kindergartener or whether you're someone else in our organization. If you have a good idea, we try to shine some light on it and then take action on it. That's one of the ways we develop leaders.

Not only do the projects expand and grow through success, but so do people. As living systems-minded leaders celebrate success and make way for others to join their efforts, they develop leaders and attract new ideas. In turn, new thinking benefits the system's efforts to move forward by countering an organization's natural pull towards equilibrium. Burke (2018) characterized equilibrium in this context as "tired thinking, solidified norms, and 'group think'" (p. 346), conditions that tend to stagnate change efforts. Infusing success awareness into the school culture, and attracting future leaders, keeps the steps towards greening a school healthy, flexible, and innovative.

Reflecting upon your community, how might promoting success benefit sustainability efforts around you? How could you communicate tangible results to those inside and outside of the organization? Does celebrating success early make you uncomfortable for any reason? If so, what beliefs need to be reframed so that you are free to confidently lead your system towards a greater vision?

Conclusion

In this last chapter, we considered four stages of gardening as a pattern for how positional leaders can take small, purposeful steps towards their vision of a vibrant, flourishing, sustainable school. Leading a school into deeper and fuller expressions of sustainability and stewardship is not a quick and simple effort. Yet, it is one that can be accomplished through a process of intentional actions. We trust that as you reach this point in the book, your commitment to facilitate change has intensified and your vision for what a school can be and do has come into greater focus. Now, by communicating vision, connecting and forming teams, considering the first seed projects to launch, and knowing where to find resources, your school will be on its way to celebrate success. Regardless of how far off the vision appears and how small the seeds currently seem, the good news is that growth and change are set in motion. By considering the idea of greening your learning environment and by reading a book such as this, you have already taken the most important step, the next one.

 Leadership Design Challenges

1. **Revisit the mission and vision**. Is it time to rewrite your mission and vision statements? Do these articulations of intention mention the truest goals your system will pursue? Spend some time reflecting on the statements of other green schools and consider how the values of stewardship could be included into your school's public professions of purpose and aspiration.
2. **Harvest successes.** What successes might you already harvest? It is likely that some seeds have actually been planted in past seasons, even if in quiet and unofficial ways. Consider where work may be taking place right now and how that work might benefit from your attention and amplification.
3. **Partner with neighbors upstream and down**. Does a district near you have a project underway? Perhaps it will work for your community as well, and so, by joining together, the planting and growing efforts become accelerated. Consider partnering with districts further along in an implementation that would also align with your priorities.

References

Burke, W. W. (2018). *Organization change: Theory and practice*. Thousand Oaks, CA: Sage Publications.

Daly, A. J., Moolenaar, N. M., Bolivar, J. M., & Burke, P. (2010). Relationships in reform: The role of teachers' social networks. *Journal of Educational Administration, 48*(3), 359–391. doi:10.1108/09578231011041062

Hawley, W. D., & Rollie, D. L. (Eds.). (2002). *The keys to effective schools: Educational reform as continuous improvement*. Thousand Oaks, CA: Corwin Press.

Kensler, L. A., Caskie, G. I., Barber, M. E., & White, G. P. (2009). The ecology of democratic learning communities: Faculty trust and continuous learning in public middle schools. *Journal of School Leadership, 19*(6), 697–735.

Kensler, L. A., & Uline, C. L. (2017). *Leadership for green schools: Sustainability for our children, our communities, and our planet*. New York: Routledge, Taylor & Francis.

Kotter, J. P. (1996). *Leading change*. Boston, MA: Harvard Business School Press.

Louis, K. S. (2007). Trust and improvement in schools. *Journal of Educational Change, 8*(1), 1–24.

Meier, D. (2002). *In schools we trust: Creating communities of learning in an era of testing and standardization*. Boston, MA: Beacon Press.

Pekarsky, D. (2007). Vision and education: Arguments, counterarguments, rejoinders. *American Journal of Education, 113*(3), 423–450.

Peterson, G., Allen, C. R., & Holling, C. S. (1998). Ecological resilience, biodiversity, and scale. *Ecosystems, 1*(1), 6–18.

Sammons, P., Hillman, J., & Mortimore, P. (1995). *Key characteristics of effective schools: A review of school effectiveness research*. Retrieved from http://files.eric.ed.gov/fulltext/ED389826.pdf

Stemler, S. E., Bebell, D., & Sonnabend, L. A. (2011). Using school mission statements for reflection and research. *Educational Administration Quarterly, 47*(2), 383–420.